Germany's Fighter Competitions of 1918

A Centennial Perspective on Great War Airplanes

Jack Herris

Great War Aviation Centennial Series #8

This book is dedicated to Carolyn Rudolph, a delightful, generous lady and wonderful friend.

Acknowledgements
My sincere thanks to Colin Owers, Aaron Weaver, Reinhard Zankl, Lance Bronnenkant, and Richard Andrews for photographs, and Greg VanWyngarden for photographs and helpful suggestions and material. Thanks to Dick Bennett for information, photos, and drawings of the Sh.III/IIIa engine. Thanks to Russell Smith for his cover painting, Bob Pearson for his color profiles, Aaron Weaver for his digital photo editing and cover design, Colin Owers and Martin Digmayer for scale drawings, and the Deutsches Technikmuseum in Berlin and the Museum of Flight in Seattle for photographs. Any errors are my responsibility.

Cover painting by Russell Smith; please see his website: **www.russellsmithart.com.** The painting, ***Mixed Company***, depicts Fokker D.VII, SSW D.III, and SSW D.IV aircraft of *Jasta* 12 on patrol together, the first two aircraft types being placed in production as a result of the First Fighter Competition and the SSW D.IV derived from the SSW D.III.

Color aircraft profiles © Bob Pearson. Purchase his CD of WWI aircraft profiles for $50 US/Canadian, 40 €, or £30, airmail postage included, via Paypal to Bob at: **bpearson@kaien.net**

For our aviation books in print and electronic format, please see our website at: **www.aeronautbooks.com**. I am looking for photographs of the less well-known German aircraft of WWI to illustrate future titles. To help with photographs please contact me at **jherris@verizon.net**.

ISBN: **978-1-935881-16-2**

Design and layout: Jack Herris
Color Profiles: Bob Pearson
Cover design: Aaron Weaver
Digital photo editing: Jack Herris & Aaron Weaver

www.aeronautbooks.com

Table of Contents

The story of the German fighter competitions of 1918 is an essential part of the larger story of late-war German fighter development. Accordingly, this book takes the opportunity briefly to tell that larger story, to put the competitions and the aircraft into the larger context of the air war that Germany, despite its technological prowess, eventually lost due to the immensely greater resources of its enemies.

Therefore some aircraft are included that were part of the development process although they did not appear at the competitions. And despite the growing importance of two-seat fighters, those are left to subsequent volumes; the subject of this volume is single-seat fighters.

Jack Herris
June 2013

Above: Albatros D.Va on display at the Australian War Memorial.

Introduction

Above: This captured Albatros D.Va is on display at the Australian War Memorial. The earlier Albatros D.I, D.II, and D.III fighters established air superiority over the Western Front in late 1916 through April 1917, but the similar D.V and D.Va were disappointing fighters generally inferior to the new generation of Allied fighters introduced in the spring and summer of 1917. Albatros fighters of these types comprised the bulk of German figther strength until the summer of 1918.

As warfare becomes more technological, small differences in technology can create large differences in combat outcomes, and aviation has generally been the area in which the greatest differences in outcomes result from the smallest differences in technology.

The German Air Service entered 1917 with optimism. Their new Halberstadt biplane fighters were the technical match of any Allied fighters and their new Albatros fighters had established a clear technical superiority over essentially all Allied combat aircraft. The major exception was the Sopwith Triplane, which demonstrated a small but important advantage in speed, climb, and maneuverability over the Albatros fighters. However, the Sopwith Triplane was a delicate airplane and most mounted only a single machine gun, whereas the Albatros fighters all mounted two synchronized machine guns, a definite advantage when gun jambs were frequent occurrences during combat. Regardless of its technical superiority, the Sopwith Triplane was not numerous enough to establish air superiority over the Western Front. However, it did cause a triplane craze at *Idflieg* that adversely affected German and Austro-Hungarian figther design throughout most of 1917.

Meanwhile, in early 1917 the German fighter units on the Western Front continued to enhance their superiority, culminating in April 1917 with a significant tactical defeat of the Allied, particularly British, air services. British losses in particular were so severe that April 1917 became known as "Bloody April" in the RFC and RNAS.

German air superiority reached its peak during April 1917 due to the technical superiority of their Albatros fighters combined with good tactics, careful use of resources, and the increasing experience, skill, and confidence of the German fighter pilots.

And as so often happens with the fortunes of war, everything was about to change. April 1917 was a turning point in the air war both strategically and tactically over the Western Front.

America Enters the War

From the strategic standpoint, the United States entered the war on the side of the Allies on April 6, 1917. While this had no immediate effect on the ground or in the air, the long-term implications were profound. The United States had huge industrial and manpower resources; in fact, these were significantly greater than Germany's, and German strength was much greater than her Austro-Hungarian and Ottoman allies. Although it would take months to create a large American army and transport it to the Western Front, there was little doubt of the eventual outcome if Germany did not win the war before the Americans arrived in strength.

The *Amerika* Program

Once America entered the war against her it was immediately clear that Germany would need to enlarge her air service to counter this new threat. In addition to enlarging the training establishment, Germany was forced to expand aircraft production to meet the challenge of America's entry into the war, and this became known as the *Amerika* Program. Key elements included:

- Increase aircraft production from 1,000 to 2,000 airplanes per month.
- Increase aero-engine production from 1,250 to 2,500 aero-engines per month.
- Increase aircraft machine-gun production to 1,500 per month.
- Increase aviation fuel output from 6,000 tons to 12,000 tons per month.
- Obtain 24,000 suitable recruits for the air service.
- Special effort started to obtain technical superiority in aircraft, especially a new fighter plane and high-performance aero-engine.

Essentially, the *Amerika* Program aimed to double the output of combat airplanes and fuel, a goal never reached. Interestingly, achieving technical superiority in aircraft, especially fighters, was a key part of the *Amerika* Program, a plan developed as German was losing her technical superiority.

New Allied Fighters Reach the Front

If the entry of America into the war had no immediate impact on the air war over the Western Front, the introduction of a new generation of Allied warplanes certainly changed the situation.

Even as German air superiority peaked in April 1917, two new British warplanes arrived that would challenge that superiority and soon reverse it. The less obvious challenger was the Bristol F2A two-seater. Originally flown like other British two-seaters, when it suffered accordingly, British crews soon learned to use it offensively as a two-seat fighter. As the crews gained experience with this offensive use, the speed, firepower, and maneuverability of the F2A gave very good results.

Similarly, introduction of the British SE5 fighter in April 1917, soon to be upgraded to the more powerful and faster SE5a, quickly showed that the new fighter out-classed the Albatros in every way. And in June, the Sopwith Camel appeared. While the Camel lacked the performance of the SE5a, it was more maneuverable than the SE5a and Albatros and was an excellent dogfighter.

The French aviation industry was similarly advancing. The Spad 7 fighter, an excellent aircraft that had suffered from limited availability due to problems with engine production, was now being delivered to the front in ever-increasing numbers. Although the Spad 7 carried only one gun, it out-performed the Albatros and was more robust.

As these new-generation warplanes were debuting over the Western Front, or finally arriving in significant numbers, the German response was the Albatros D.V, a lightened, more streamlined derivative of the D.III already in service. Using the same 160 hp Mercedes D.III engine as the earlier Albatros fighters, the D.V offered little or no performance improvement. Worse, lightening its structure weakened its airframe. Failures of its single-spar lower wing during combat, already problematic in the Albatros D.III, became a serious difficulty for the D.V, further limiting its effectiveness.

Yet another problem for the German air service was the Albatros D.V had been ordered into large-scale production as part of the *Amerika* Program because no superior fighters were available. The Halberstadt fighters were robust, maneuverable aircraft with excellent flying qualities, but the Albatros out-performed them. The Roland fighters were strong, fast, and heavily-armed like the Albatros, but had poor maneuverability and handling qualities and were no match for the new Allied fighters. And contemporary Fokker biplanes were fragile and inferior in performance and therefore had already been removed from the front.

Through April the Albatros fighters swept the skies of the Western Front of Allied warplanes, but by July the Red Baron, by then the leading German ace, was complaining about the 'wretched Albatros' that he and his men had to fly because there was no better German alternative. The new Pfalz D.III that arrived in August was not a solution; although more robust than the Albatros, its performance and maneuverability were similar. How could Germany again achieve the needed technical superiority?

Allied Fighter Competitors

There were other, less capable Allied fighters such as the Nieuports and DH-5 that were not nearly as good as these fighters, but there were many of the best types shown here, all of which were superior to the Albatros and Pfalz in late1917/early 1918.

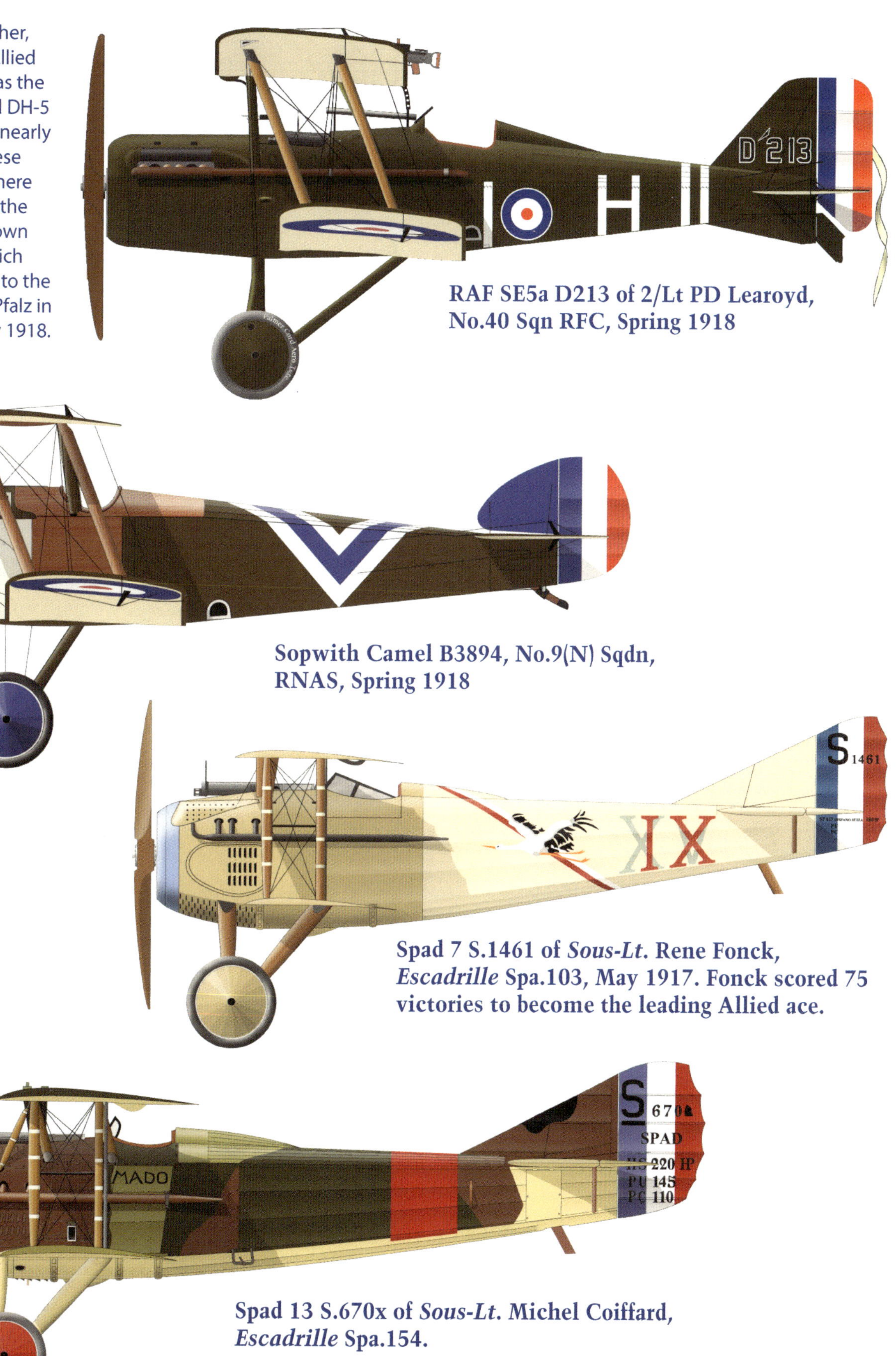

RAF SE5a D213 of 2/Lt PD Learoyd, No.40 Sqn RFC, Spring 1918

Sopwith Camel B3894, No.9(N) Sqdn, RNAS, Spring 1918

Spad 7 S.1461 of *Sous-Lt*. Rene Fonck, *Escadrille* Spa.103, May 1917. Fonck scored 75 victories to become the leading Allied ace.

Spad 13 S.670x of *Sous-Lt*. Michel Coiffard, *Escadrille* Spa.154.

The Triplane Craze

One possibiliy was the triplane. During the period of German resurgence over the Western Front in late 1916 and early 1917, two important new Allied fighters were introduced in small numbers. One was the Spad 7 biplane. Powered by the innovative new Hispano-Suiza V-8, the Spad was strong, fast, and maneuverable. Its main limitation as a fighter was that it only carried one synchronized machine gun. Furthermore, engine production problems initially limited its numbers, and therefore its impact.

While the French introduced the sturdy Spad biplane, the British RNAS introduced a more distinctive fighter, a triplane development of the agile Pup fighter. The Sopwith Triplane retained the single gun and general lines of the Pup, but had a more powerful engine coupled with a new triplane wing cellule.

Apparently the Sopwith Triplane's configuration was the key to its impact on *Idflieg*. While a good fighter, the Spad was a biplane and its performance could be attributed to its innovative engine – which is where *Idflieg* should have focused more attention. On the other hand, the Sopwith Triplane was the first triplane fighter to appear over the front, and its rotary engine was a known quantity. In any case the Sopwith Triplane caught the fancy of *Idflieg*, who imagined its triplane configuration was the secret to its success. The result was the Triplane Craze, an affliction confined almost entirely to Germany and Austria. German authorities, followed by the Austrians, requested their manufacturers develop triplane fighters, resulting in a plethora of new triplanes that never reached the front and wasted precious engineering resources in the process. Of these designs only the Fokker Triplane, made famous as the airplane the Red Baron was flying when he was killed, saw combat in significant numbers.

The Fokker Triplane came at a time when the Albatros, Roland, and Pfalz fighters were inferior to their Allied opponents, so it was welcomed warmly by the German pilots and became a favorite mount of German aces despite being slow, its main limitation. On the positive side, it had exceptional climb and maneuverability ideally suited to close-in, high-G maneuvering combat, or dog-fighting as it was called, that was so characteristic of WWI air combat over the Western Front. While it needed more speed to be a first-class fighter, at least it out-performed its opponents in climb and maneuverability, whereas the Albatros, the most numerous fighter, was now inferior in all respects.

The German Triplane Craze is worthy of its own story, documented in Volume 7 of this series. However, eventually reality prevailed, and designers abandoned the triplane configuration due to its higher drag.

To its credit *Idflieg* did order development of V-8 fighter engines, but none reached production in time to power a German fighter in combat.

Below: A sight to stir aviation enthusiasts everywhere, flying replica Fokker Triplanes in authentic colors zoom through calm New Zealand skies during an airshow. The charismatic Fokker Triplane remains popular a century after its appearance in the skies over the Western Front.

German Fighters Over the Front

To keep this discussion in perspective, the inventory of actual fighters at the front is helpful. The *Frontbestand* tables are the actual front-line inventories of the various classes of aircraft by type every other month, and here we show the three tables covering fighters. The *Frontbestand* table for E-types shows monoplane fighters, that for D-types shows biplane fighters, and that for Dr-types shows triplane fighters. The D-type biplane fighters were clearly far more numerous, and in September 1918 *Idflieg* decided to use the D-type designation for all fighter configurations, rendering the E-type and Dr-type designations obsolete after that time.

By the end of 1916 monoplane fighters virtually disappeared from the front with the exception of a few training machines until the brief appearance of the Fokker E.V in August 1918. The E.V was removed from the front due to wing failures, and when the fighter re-appeared in October with replacement wings it was designated the D.VIII.

The story of German triplane fighters at the front is basically the story of the Fokker Triplane, which despite its fame was never very numerous. Thus the table of greatest relevance to our story is the D-type *Frontbestand* table for biplane fighters at the front.

The D-type inventory shows the great dominance of the Albatros fighters from late 1916 into early 1918. However, the Fokker D.VII replaced the Albatros as the main front-line fighter as quickly as it could be produced, and was the most numerous German fighter during the last months of the war.

Right: This reproduction Albatros D.V is in authentic camouflage and markings. The Albatros D.V and D.Va were the mainstay German fighters from the second half of 1917 until replaced by the Fokker D.VII in the summer of 1918.

Frontbeststand Inventory of Dr-Type Aircraft (Triplane Fighters) at the Front

Manufacturer and Type		1914 31 Aug	1914 31 Oct	1914 31 Dec	1915 28 Feb	1915 30 Apr	1915 30 Jun	1915 31 Aug	1915 31 Oct	1915 31 Dec	1916 28 Feb	1916 30 Apr	1916 30 Jun	1916 31 Aug	1916 31 Oct	1916 31 Dec	1917 28 Feb	1917 30 Apr	1917 30 Jun	1917 31 Aug	1917 31 Oct	1917 31 Dec	1918 28 Feb	1918 30 Apr	1918 30 Jun	1918 31 Aug
Fokker	Dr.I																			2	17	35	143	171	118	65
Pfalz	Dr.I																							9	7	1
Total:																				2	17	35	143	180	125	66

Frontbeststand Inventory of E-Type Aircraft (Monoplane Fighters) at the Front

Manufacturer and Type		1914 31 Aug	1914 31 Oct	1914 31 Dec	1915 28 Feb	1915 30 Apr	1915 30 Jun	1915 31 Aug	1915 31 Oct	1915 31 Dec	1916 28 Feb	1916 30 Apr	1916 30 Jun	1916 31 Aug	1916 31 Oct	1916 31 Dec	1917 28 Feb	1917 30 Apr	1917 30 Jun	1917 31 Aug	1917 31 Oct	1917 31 Dec	1918 28 Feb	1918 30 Apr	1918 30 Jun	1918 31 Aug
Fokker	E.I						4	17	23	26	12	22	3	1												
	E.II							2	8	14	20	13	4	5	2	1					2					
	E.III							3	23	40	67	110	101	64	28	11	2	1	1		8	7				
	E.IV								1	6	6	28	29	25	16	2										
	E.V (D.VIII)																									80
Pfalz	E.I								2	14	26	27	13	11	3	1										
	E.II									6	11	20	30	17	3											
	E.III								1	1		4	8													
	E.IV											5		1												
	E.V												3													
SSW	E.I														5	2	1	1								
Unknown				2																						
Total:				2			4	22	58	107	142	229	191	124	57	17	3	2	1		10	7				80

Frontbeststand Inventory of D-Type Aircraft (Biplane Fighters) at the Front

Manufacturer and Type		1914			1915						1916						1917						1918			
		31 Aug	31 Oct	31 Dec	28 Feb	30 Apr	30 Jun	31 Aug	31 Oct	31 Dec	28 Feb	30 Apr	30 Jun	31 Aug	31 Oct	31 Dec	28 Feb	30 Apr	30 Jun	31 Aug	31 Oct	31 Dec	28 Feb	30 Apr	30 Jun	31 Aug
AEG	D.I																			1						
Albatros	D.I													1	50	39	28	20	17	12	9	8	5	6	1	3
	D.II													1	28	214	150	107	72	44	11	6	2	2	2	2
	D.III															13	137	327	303	385	446	423	357	174	82	52
	D.V																		216	424	526	513	250	131	91	20
	D.Va																				53	186	475	928	604	307
	D Lf																	2								
Aviatik	D.I																	15								
	D.III																	4								
DFW	D.II																		1							
Euler	D.II														2							11	16	13	6	1
Fokker	D.I													10	74	4	5	5	6	2	1					
	D.II													10	49	68	49	33	26	10			6			
	D.III													7	6	34	7			1		4		11	11	10
	D.IV													2		2					4					
	D.V																3	6	4	4	4	12	25	19	6	3
	D.VI																								21	27
	D.VII																							19	407	828
Halberstadt	D.I														6		12	16	14	1	2					
	D.II												6	5		55	5	2	14	16	3	2	1		1	1
	D.III										1		2	20	32	17	12	9	8	6	7	5	1	2	2	2
	D.V														17	32	39	11	6	9	5	3	2			
LVG	D.I																61	47	18							
	D.II																1									
Pfalz	D.III																			3	145	276	182	13	13	3
	D.IIIa																					114	261	433	289	166
	D.VIII																								14	19
	D.XII																								5	168
Roland	D.I														1	7	12	2	1							
	D.II																22	97	41	10	2					
	D.IIa																		128	102	6	3				
	D.III																						9	14		
	D.VIa																								38	58
	D.VIb																								17	12
SSW	D.I																	2								
	D.III																							4	4	6
	D.IV																									3
Total:											1		8	62	265	485	562	686	875	1030	1224	1566	1592	1769	1614	1691

D-Type Frontbestand Notes

The Aviatik D.I and D.III listed in February 1917 were the Halberstadt D.II(Av) built under license by Aviatik. The DFW D.II listed in June 1917 may be one of the DFW *Floh* (flea) prototypes then under development that may have been sent to the front for evaluation.

The Euler D.II entry for October 1916 must be an error; it was type-tested on 8 January 1917 prior to an order for 30 production fighters placed in March 1917. The unarmed Euler D.II served as a transition trainer to familiarize pilots with rotary engine aircraft. Similarly, the build-up of Fokker D.V rotary-powered fighters in late 1917 and early 1918 was due to moving these fighters near the front for training pilots transitioning to the rotary-powered Fokker Dr.I.

The LVG D.I was the Albatros D.II(LVG) built under license and should have been included in the Albatros D.II entries. The LVG D.II was the LVG D 12 biplane fighter prototype developed in late 1916 and, if this entry is not an error, sent to the front for evaluation.

Upon its arrival at the front the Fokker D.VII became the dominant fighter and its numbers grew as quickly as production allowed.

The German Fighter Competitions

For a number of reasons – technical stagnation, partly due to engine development stagnation, increasing Allied technical and numerical superiority, and American entry into the war making all these problems worse – German fighter aviation was under great stress in the second half of 1917.

At this time, Manfred von Richthofen, the 'Red Baron', Germany's leading ace famous throughout the Empire, contacted his friend *Oblt.* Fritz von Falkenhayn, (son of *Gen.* Erich von Falkenhayn, Chief of the General Staff from Sept. 1914 to August 1916) who was *Kogenluft's* adjutant 1915–1918, and complained strongly about 'the wretched Albatros' fighters. Richthofen suggested a fighter competition between all the German manufacturers to develop a superior new fighter. Such was his influence, and the common sense of the solution he recommended, that *Idflieg* decided to follow his suggestion, and the first German fighter competition was held 20 January–12 February 1918.

Above: Manfred von Richthofen, the famous Red Baron, was not just the war's leading ace and fighter leader. He also served his country by instigating the fighter competitions held three times during 1918 that were instrumental in getting new fighter types to the front.

All the manufacturers entertained the pilots and officials to make them well-disposed to their aircraft. However, *Idflieg* had given a lot of thought to the fighter competition process and designed it to be as fair and objective as possible and to address the priorities of successful fighter pilots at the front.

To ensure unbiased results, strict procedures were instituted. Once the prototypes were handed over to the Adlershof staff, they were basically in bond, and no modifications or adjustments were to be made by the manufacturers. The prototypes were checked and weighed before each flight. Details of the propellers fitted were recorded, and the fuel consumed during each flight was recorded. Two calibrated and sealed barographs were fitted to each aircraft; these were checked before and after every flight.

At the beginning of the trials the aircraft were demonstrated by company pilots under supervision of the Adlershof staff. The second part of the trials were flown by Adlershof pilots and front-line pilots from the *Jastas*, especially well-known, successful aces. The front-line pilots were to fly the prototypes for the key part of the trials, including climb and speed performance, maneuverability, and simulated combat with other prototypes and service aircraft provided for comparison. These aces were to evaluate the competitors and make recommendations about which designs should actually be produced. In addition to testing flight performance, the aircraft were evaluated for strength, reliability, and maintainability.

Moreover, examples of the fighters currently at the front, or more powerful versions of them, were included in all the competitions to provide a practical standard of comparison for the competitors.

The key result of the First Fighter Competition, whose formal trails were held from 21 January until 12 February, 1918, was the selection of a superior fighter, the Fokker D.VII, to replace the outdated Albatros and Pfalz fighters. Significantly, the D.VII was initially powered by the same Mercedes engine used in those fighters, but soon a new engine, the over-compressed 185 hp BMW.IIIa became available, enabling more of the airframe's potential to be exploited. Mercedes responded with improved

Above: The most important immediate result of the First Fighter Competition, and arguably the most important result of all the fighter competitions, was the selection of the Fokker D.VII for large-scale production for the German fighter units. This fighter regained technical superiority for Germany and enabled her pilots to double their scoring rate. The photo is an original D.VII in its original camouflage now on display in Knowlton, Canada.

versions of their basic engine offering similar performance. The resulting BMW-powered D.VII fighter is thought by many, including the author, to be the finest all-around fighter of the war.

The other important result of the First Fighter Competition was that it showed the efficacy of the process in procurring new, superior fighters for the air service. In general, the fighter competitions provided more structure, focus, and momentum to the German effort to develop and procure better fighters for their air service. Therefore, two subsequent fighter competitions were held.

In mid May all manufacturers were notified that their prototypes had to be at Adlershof by 31 May. The aircraft were first inspected, weighed, and prepared for testing. Performance testing was done June 1–21, followed by comparative evaluation.

The Second Fighter Competition resulted in production of the Fokker E.V/D.VIII, a parasol-monoplane development of the Fokker Dr.I and D.VI. This innovative aircraft had a troubled history due to manufacturing quality problems, but the design itself was sound and offered excellent performance considering its simplicity and low power.

The Third Fighter Competition, officially restricted to fighters with BMW.IIIa engines but including rotary-powered types for comparison, was held October 15–31, 1918. The Fokker V.29, a parasol-monoplane derivative of the Fokker D.VII, was one of two winners and may have gone into production except for the Armistice. The Rumpler D.I, also chosen as a winner, was still not ready for production before the Armistice despite a prolonged development process that had started in early 1917. Whether the troubled Rumpler D.I would ever have been brought to production-ready status as a practical fighter is doubtful.

Given the results of the competitions, some critics have contended that Anthony Fokker had an unfair advantage. Certainly Fokker himself was an exceptional pilot, and he also had great rapport with many of the fighter pilots evaluating the prototypes at the competitions. This clearly gave him greater insight into the real needs of the combat pilots at the front than his competitors, and his abilities as a test pilot helped him fine-tune his aircraft to meet the needs of the aces. These were certainly advantages, but not unfair ones; Fokker's aircraft still had to deliver better performance than those of the other manufacturers to win the competitions. Moreover, from the sheer number of different prototypes he brought to the competitions, it is apparent that Fokker simply worked harder to succeed than the other manufacturers.

Aces at the Fighter Competitions

In addition to factory and *Idflieg* pilots, successful aces were brought back from the front to evaluate the prototype fighters. Whereas test pilots could evaluate the fighters' performance objectively in terms of speed and climb rate, the less objective characteristics that are essential to a fighter, maneuverability and flying qualities, were to be judged by the men who knew most about them – front-line aces who depended on their aircraft daily and had mastered the art of combat flying.

The First Fighter Competition

Front-line combat pilots known to have attended the First Fighter Competition were *Rittmeister* Manfred *Freiherr* von Richthofen (80 victories, *JG*I), *Oblt.* Hans Klein (22 victories, *Jasta* 10), *Hptm.* Bruno Loerzer (44 victories, *Jasta* 26) *Lt.* Erich Loewenhardt (54 victories, *Jasta* 10), *Hptm.* Adolph *Ritter* von Tutschek (27 victories, *JG*II), *Lt.* Josef Jacobs (48 victories, *Jasta* 7), *Lt.* Joachim von Busse (11 victories, *Jasta* 20), *Lt.* Heinrich Geigl (13

Above: Taken at the First Fighter Competition, this photograph shows, from left, *Lt.* Erich Loewenhardt, *Hptm.* Bruno Loerzer, *Rittmeister* Manfred von Richthofen, *Hptm.* Curt Schwarzenberger (the *Referent* for fighter aircraft development at *Idflieg*), *Oblt.* Hans Klein, and unknown.

Right: Probably taken at the Third Fighter Competition, this photograph shows Ernst Udet (62 victories) at left and Josef Jacobs (48 victories) at right. Both were awarded the *Pour le Mérite;* Udet's is clearly visible and Jacob's is barely visible. Both survived the war.

Above: Taken at the Second Fighter Competition, this photograph shows Friedrich Seekatz of the Fokker company at far left. From left to right, the pilots are *JG*III commander Bruno Loerzer, *Oblt*. Hermann Göring (soon to command *JG*I), *Lt*. Lothar von Richthofen, *Lt*. Hans Kirschstein (*Jasta* 6), *Lt.d.R.* Constantin Krefft (*JG*I Technical Officer), *Lt*. Friedrich Mallinckrodt (formerly of *Jasta* 6, now with *Idflieg*), and *Lt*. Fritz Schubert (*Jasta* 6 and also on the technical staff of *JG*I.)

Left: Taken at the Third Fighter Competition in October, this photograph shows Alfred Eversbusch of the Pfalz company standing in the middle surrounded by aces. In the front row were the three *Jagdgeschwader* commanders, from left *Oblt*. Hermann Göring (*JG*I), *Hptm*. Bruno Loerzer (*JG*III), and *Oblt*. Oskar *Freiherr* von Boenigk (*JG*II). From left in the back row, *Lt*. Hans Klein, *Lt*. Josef Veltjens, Ernst Eversbusch, *Lt*. Ernst Udet, and *Lt*. Josef Jacobs. Von Boenigk, the only ace not wearing a *Pour le Mérite,* would be awarded the *Pour le Mérite* during the competition.

victories, *Jasta* 16), *Oblt.z.S.* Gotthard Sachsenberg (31 victories, *MFJ*I), and *Oblt.z.S.* Theo Osterkamp (32 victories, *MFJ*I).

Other evaluation pilots at the First Fighter Competition included *Freiherr* von Esebeck, Grieffenhagen von der Osten, and Hans Rolshoven.

The Second Fighter Competition

Front-line pilots known to have participated in the Second Fighter Competition included *JG*III commander Bruno Loerzer (44 victories), *Hptm.* Wilhelm Reinhard (*JG*I commander, 20 victories), *Oblt.* Hermann Göring (who would command *JG*I after Reinhard's fatal accident at the evaluation, 22 victories), *Lt.* Lothar von Richthofen (40 victories, *Lt.* Hans Kirschstein (*Jasta* 6, 27 victories), and *Fw.Lt.* Fritz Schubert (of *Jasta* 6 and later on the technical staff of *JG*I, 3 victories).

Lt. Friedrich Mallinckrodt, now with *Idflieg,* was formerly of *Jasta* 6 (6 victories), and *Lt.d.R.* Constantin Krefft, also now with *Idflieg,* was formerly *JG*I Technical Officer (2 victories). *Hptm.* Schwarzenberger was another evaluation pilot.

The Third Fighter Competition

Front-line pilots known to have participated in the Third Fighter Competition included *Lt.* Paul Bäumer (43 victories, *Jasta* Boelcke), *Lt.* Walter Blume (28 victories, *Jasta* 9), *Lt.* Heinrich Bongartz (33 victories, *Jasta* 36), *Oblt.* Oskar *Freiherr* von Boenigk (26 victories, *JG* II), *Lt.* Franz Büchner (40 victories, *Jasta* 13), *Lt.* Walter Dingel (2 victories, *Jasta* 15 and staff *JG*II), *Lt.* Josef Jacobs (48 victories, *Jasta* 7), *Lt.* Hans Klein (22 victories, *Jasta* 10), *Hptm.* Bruno Loerzer (44 victories, *JG* III), *Lt.* Eberhardt Mohnicke (9 victories, *Jasta* 11), *Lt.* Franz Ray (17 victories, *Jasta* 49), *Lt.* Lothar von Richthofen (40 victories, *Jasta* 11), *Lt.* Hans-Joachim Rolfes (17 victories, *Jasta* 45), *Hptm.* Eduard *Ritter* von Schleich (35 victories, *JG* IV), *Lt.* Emil Thuy (35 victories, *Jasta* 28), *Lt.* Ernst Udet (62 victories, *Jasta* 4), and *Lt.* Joseph Veltjens (35 victories, *Jasta* 15).

In addition, *Lt.* Harry von Bülow-Bothkamp (6 victories) had been a front-line pilot and was now assigned to *Idflieg. Lt.* Friedrich Mallinckrodt (6 victories) was another former front-line pilot also now assigned to *Idflieg*). *Oblt.* Fritz von Falkenhayn, *Kogenluft's* adjutant, also attended.

The front-line aces who participated in the fighter competition evaluations were highly successful fighter pilots and most, but not all, were either holders of the *Pour le Mérite* or soon would be.

Above: Probably taken at the Third Fighter Competition, this photograph shows *Hptm.* Bruno Loerzer (44 victories, *JG*III) at left, Anthony Fokker in the middle, and *Oblt.* Hermann Göring (22 victories, *JG*I) at right.

Above: Taken at the Second or Third Fighter Competition, this photograph shows *Oblt.* Hermann Göring (22 victories, *JG*I) at left, Friedrich Seekatz of the Fokker company in the middle, and *Hptm.* Bruno Loerzer (44 victories, *JG*III) at right.

Above: If *Idflieg* was determined to ensure unbiased results during the 1918 Fighter Competitions in Berlin, the aircraft manufacturers were equally determined to ensure the visiting fighter aces were favorably disposed toward them and, by extension, their products. Fokker, Albatros, and Pfalz all generously wined and dined the aces. Here in October 1918 at the lavish Hotel Adlon are (left to right bottom row): *Oberleutnant* Oskar *Freiherr* von Boenigk, *Leutnant* Walter Dingel, *Leutnant* Paul Bäumer, *Leutnant* Franz Büchner, and *Leutnant* Hans Joachim Rolfes; (middle row) *Leutnant* Josef Veltjens, *Leutnant* Hans Klein, *Oberleutnant* Hermann Göring, *Hauptmann* Bruno Loerzer, *Oberleutnant* Ernst Udet, *Leutnant* Josef Jacobs; (standing) Pfalz *Direktor* Hiehle, Pfalz *Direktor* Ernst Eversbusch, Pfalz *Direktor* Robert Kahn, *Leutnant* Friedrich Mallinckrodt.

Right: Pfalz made an especially dramatic impression with their entertainment, persuading noted Viennese dancer Lucy Kieselhausen – at great expense – to dance nude for the thunderstruck pilots! Lucy evolved out of the ballet culture and favored luxuriously decorative hothouse costumes and the utmost refinement of movement. "For Lucy the waltz was a distillation of the stirrings within an opulent boudoir, with its background of privileges and secrets… Her joyfully flashing temperament did not hover on a smooth surface but over a shadowy abyss from which issued her fool's dance with its slumbering, half-animal rapture." Lucy (1897–1927) went on to roles in film after the war, but died young in a tragic accident.

Above: Like Fokker and Albatros, the Pfalz company spared little expense to entertain the front-line pilots who evalutated the prototype fighters and recommended which should be produced. Here *Lt.* Josef Veltjens and other airmen are attending a dinner hosted by Alfred Eversbusch (Pfalz Company) during the Third Fighter Competition at Adlershof. Eversbusch sits at far right, and next to him are: *Hptm.* Bruno Loerzer, *Oblt.* Ernst Udet, *Hptm.* Eduard Ritter von Schleich, *Oblt.* Oskar von Boenigk, *Lt.* Josef Jacobs, unknown, *Lt.* Franz Ray and other unknowns. Across the table, starting at the right, are: *Oblt.* Hermann Göring, *Lt.* Friedrich Mallinckrodt, *Lt.* Heinrich Bongartz, *Lt.* Hans Klein, *Oblt.* Fritz von Falkenhayn, Veltjens, unknown, *Lt.* Paul Bäumer, *Lt.* Hans-Joachim Rolfes, *Lt.* Walter Dingel, unknown. (courtesy Lance Bronnenkant)

Right: Lucy Kieselhausen in a less elaborate costume than she normally favored. Even if available, a photo of her in her 'costume' for the entertainment she provided the aces for the benefit of Pfalz could not be included in this book!

Above: Rotary-powered fighter prototypes lined up, apparently at the Second Fighter Competition. First in line is a prototype Pfalz D.VIII with enlarged cowling, N-struts, and horn-balanced ailerons, all features not found on the production model. Next in line appears to be a Roland D.XIV prototype, then a Roland D.IX prototype. Fourth in line is an SSW D.III. Photographers are lined up to take portraits, while the armed guard at left seems bored. (via Reinhard Zankl)

Below: A Fokker Dr.I shown on a snowy airfield reproduced as Sanke Card 1055. The Fokker Dr.I was in production in January 1918 to equip the *Jastas* with a new fighter for the coming spring offensive, and two were at the competition for comparison with the new fighters being evaluated. The production Dr.I used the 110 hp Oberursel Ur.II engine, but one evaluation Dr.I had the 145 hp Oberursel Ur.III and the other had the 160 hp Goebel Goe.III. Both engines offered usefully more power than the standard Ur.II, but neither was in production because more development was needed.

Fighter Engine Development in 1918

German fighter development in 1918 was paced by the availability of new, more powerful fighter engines. The engine is critical because it determines the aircraft's overall performance potential. For example, the Fokker D.VI with 110 hp Ur.II was a good fighter, with excellent maneuverability and handling quailities and good speed and climb at lower altitudes; with more power it could have been a great fighter. It was tested with 145–160 hp engines, but these were not available for production.

The 1918 German fighter engine story has three main threads; the inline six-cylinder engines, V-8 engines, and high-power rotary engines.

The reliable 160 hp Mercedes D.III had now powered aircraft at the front for nearly two years, and while it was an excellent engine, the fighters now needed more power for higher ceiling, better rate of climb, and higher speed. The slightly higher-compression Mercedes D.IIIa, which typically gave 170 hp, introduced in mid 1917 was a step in the right direction but more was needed.

Fortunately, help was at hand with the new, over-compressed 185 hp BMW.IIIa. Like the Mercedes an inline six, the BMW was designed to be run at full throttle at 2,000 meters and above. Below that altitude it had to be run at reduced throttle setting to avoid detonation and severe engine damage. In contrast to the Mercedes D.IIIa that gave 170 hp at sea level, the BMW gave 185 hp at 2,000 meters. The BMW provided excellent high-altitude performance and reliability; there just were not enough of them.

The debut of the BMW was the motivation that Daimler needed to improve their Mercedes D.IIIa, and new, over-compressed versions of that engine, the D.IIIaü and D.IIIavü, gave up to 220 hp and matched the BMW engine's power at high altitude. So the inline-six story was reasonably good; the engines were reliable and offered good performance, but production quantity never met the needs of the front. And as always, more power was wanted.

In fact, what was needed was a high-performance V-8 engine comparable to the excellent Hispano-Suiza used in Spads, the S.E.5a, and the Sopwith Dolphin. Here both Daimler and Benz offered V-8 prototypes that failed to achieve production status soon enough to reach combat during the war despite the advantage of having captured Hispano-Suiza engines as examples. The Mercedes D.IIIb V-8 and Benz Bz.IIIb V-8 powered various fighter prototypes flown at the fighter competitions but never powered a fighter in combat, a serious setback for German fighter aviation. Moreover, these engines were designed to achieve 185–195 hp at a time when the contemporary Hispano-Suiza V-8 was available in 200 hp, 220 hp, and 235 hp versions, and the new engine moving toward production was producing 300 hp. Unaccountably, the Germans had taken dead aim at a moving target. So the V-8 part of the fighter engine story was one of failure.

The German rotary-engine story was a little brighter, but still not fully successful. While several companies developed high-power rotary engines, only the Siemens-Halske engines powered fighters in combat. The 11-cylinder, counter-rotary 160 hp Sh.III/IIIa gave exceptional performance, especially in climb, but development was protracted and production was too limited.

Oberursel and Goebel produced prototype high-power rotary engines but could not develop them into reliable production engines in time for combat. Part of the problem for all rotary engines was Germany's lack of a reliable lubricant due to the Allied distant blockade. So the high-power rotary engine story was one of only partial success, highlighted by failure to produce enough reliable engines and deliver them soon enough.

Right: The Pfalz D.VII was one of the prototypes at the First Fighter Competition powered by the 160 hp Sh.III counter-rotary. For adequate cooling the cowling had to be extensively cut away and numerous cooling holes added. The Sh.III gave good speed and an exceptional rate of climb, but required prolonged development.

The Siemens-Halske Sh.III Engine

by Dick Bennett

Of Germany's last-generation rotary engines, only the Siemens-Halske Sh.III saw more than token service. Unlike conventional rotaries, where the crankshaft was fixed to the airframe and only the cylinders and propeller rotated, the Sh.III was a counter-rotary design, where the crankshaft turned in one direction, and the cylinders and propeller rotated together in the opposite direction. In essence, it combined the features of a traditional rotary engine and a modern radial.

Siemens believed this mechanical complexity had several benefits:

1. Maximum crankshaft speed, relative to the crankcase-cylinder assembly, was 1800 rpm, vs. 1200–1300 for fixed-crankshaft rotaries. An engine of a given displacement would develop higher horsepower.
2. Torque effect, which became stronger as engine power increased, was significantly reduced by the lower rotational speed of the cylinders and further offset by the torque effect of the internal parts turning in the opposite direction.
3. The lower its rpm, the more efficient the propeller at developing thrust.
4. The lower rotational speed of the engine led to less air friction power loss.

The Sh.III was an 11-cylinder, single-row engine, with a differential gear case, anchored to the airframe, mounted at its rear. Power from the rotating crankshaft was transmitted by this gear train to the crankcase and cylinders, driving them in the opposite directions. Pushrod-operated intake and exhaust valves were located in the cylinder heads. Except for the cast aluminum pistons, all major components were steel.

At 5.1:1, the Sh.III was considered a high compression engine. By today's standards, that seems comically low, but this was the day and age of aero engine compression ratios between 4 and 4.5. The engine developed 160 hp on initial testing in June 1917, but with continuing refinement, eventually produced over 200 hp at sea level.

The engine's performance and potential for high altitude use caught the attention of *Idflieg*, but the clincher was the spectacular climbing performance of an Sh.III-powered SSW D.III prototype in September 1917. *Idflieg* immediately ordered a limited production run, in advance of the customary endurance test. Eventually, Siemens-Halske received

Left: Front view of the Siemens-Halske Sh.III. The eleven-cylinder engine had a bore of 124mm, a stroke of 140mm, and a displacement of 18.60 liters. Originally delivering 160 hp, it was gradually developed to deliver 220–240 hp. Unexpectedly, the engines produced under license by Rhemag were more reliable than those produced by Siemens-Halske. Fighters using the Sh.III/IIIa had good speed and an exceptional rate of climb.

an order for 2,000 engines, and a manufacturing license and order for 1,000 more were later placed with Rhemag (Rhenania Motorenfabrik A.G.).

On the basis of the January Fighter Trials, *Idflieg* ordered trial production runs of SSW D.III and Pfalz D.VIII fighters powered by the Sh.III. The first examples arrived at the front in March and April 1918. After a few hours' operation, the engines began seizing, with pistons, rings, valves and other components failing. Engine production was stopped; the surviving aircraft were returned to the factories.

There were plenty of suspects – inadequate cooling, the castor oil substitute used by the engines, component design, and metallurgy. In a crisis – Germany was at war and there was no time for a careful evaluation – everything was "fixed", whether necessary or not.

Most changes involved engine cooling. The bottom half of the SSW D.III cowling was cut away, scoops were added to the propeller spinner to direct air onto the crankcase, and an additional exhaust duct was shoehorned into the front of the fuselage, its outlet discharging on the right side.

The original pistons had a rather deep, square-shouldered recess in the crowns, and designers suspected its sharp, drastic section changes might be overstressing the castings. Consequently, they redesigned them with a shallower dished recess.

This created another problem. Because the piston recess was part of the combustion chamber, reducing its volume caused a corresponding reduction in combustion chamber volume, and with it, power, but it increased the compression ratio on an already troubled engine. Siemens-Halske's elegant solution was to add 10 mm to the cylinder length. This restored the combustion chamber volume and compression ratio, allowing existing crankshafts and crankcases to be reused. Engine diameter increased from 1050 to 1070 mm, but the only other external difference was two additional fins on each cylinder.

The modified engine passed *Idflieg's* endurance test and was released for production as the Sh.IIIa, to emphasize it wasn't the trouble-plagued original. It was not a higher compression or more powerful version of the Sh.III – all the changes had simply maintained its original performance.

Siemens-Halske and Rhemag produced between 800 and 840 Sh.IIIa engines by the Armistice, but no more than 250 of them found their way to the front in SSW D.IIIs and D.IVs and Pfalz D.VIIIs, where the modified engines performed reliably.

Right: Rear view of the Siemens-Halske Sh.III eleven-cylinder counter-rotary engine. The low rotational speed of the cylinders, 900 RPM compared to 1200–1300 RPM for conventional rotary engines, reduced the cooling air flow over the cylinders, which reduced cooling drag but also made cooling marginal. Problems with insufficient cooling caused a number of engine changes as well as airframe changes to the fighters using this engine to increase cooling air flow over the cylinders. These airframe changes were especially noticeable in the SSW D.III and D.IV fighters, which had a number of engine cowling variations as well as the addition of cooling air scoops to the propeller spinner.

The First Fighter Competition

The First Fighter Competition was held from 20 January to 12 February 1918. Significantly, only one of the *new* designs submitted was a triplane. Two Fokker Triplanes, an aircraft then in production, were tested. However, the inherent weight and drag limitations of the triplane configuration were finally being recognized and after the First Fighter Competition no new triplane fighter designs were ever submitted to the Fighter Competitions. One retrograde *quadraplane* design, the Naglo D.II, was submitted for flight evaluation at the Second Fighter Competition, but no record of its performance is available. In fact, several monoplanes were submitted to the First Fighter Competition. The Triplane Craze instigated by *Idflieg* in early 1917 was finally coming to a close, and monoplane fighter prototypes became increasingly common during this last year of the war.

Examples of the main production fighters then at the front, the Albatros D.Va, Pfalz D.IIIa, and Fokker Dr.I, were entered in the competition to provide a standard for comparison with the new designs. Two Albatros D.Va fighters and both the Pfalz D.IIIa fighters at the competition had their standard engines. In addition, two more Albatros D.Va fighters had new, high-compression engines. Both Triplanes at the competition used more powerful engines than the standard production aircraft.

Although the competition was open to all designs, a winner was to be declared in two categories, one for aircraft powered by the existing 160 hp Mercedes D.IIIa engine and one for rotary-powered fighters. In the Mercedes-engine category the modified Fokker V11 was selected for production as the Fokker D.VII. The Roland D.VIa was runner up, and an evaluation batch of 50 Roland D.VIa fighters was also ordered. Simultaneously, the semi-autonomous Bavarian authorities chose the Pfalz D.XII to replace the D.IIIa in production despite the fact that the Pfalz D.XII was not present at the competition. Not coincidentally, the Pfalz company was located in Bavaria and patronized by the Bavarian government.

The Aviatik D.III, powered by a pre-production Benz Bz.IIIb water-cooled V-8 engine, was also present, and the new BMW.IIIa over-compressed straight-six engine made its debut in an Albatros D.Va, as did the Mercedes D.IIIaü over-compressed engine. These engines improved the climb rate of the Albatros but could not solve the inherent limitations of its airframe, especially its weak lower wing spar.

In the rotary-engine category the Fokker D.VI, powered by the readily-available but low power 110 hp Oberursel Ur.II, was chosen as one winner. Additionally, there was a new rotary engine in development, the innovative 160 hp Siemens-Halske Sh.III counter-rotary. Offering much more power than the Oberursel Ur.II, the Sh.III powered prototypes from SSW Pfalz, Fokker, and Roland, and evaluation batches of both the SSW D.III and Pfalz D.VII/VIII were ordered. In the event, the stronger, two-bay D.VIII was chosen for production in preference to the single-bay D.VII.

The presence of fighters powered by new, pre-production engines in addition to the long-time standard 160 hp Mercedes D.IIIa was another encouraging sign at the competition. Unfortunately, only the BMW and Seimens-Halske engines were to appear in combat; the other new engines did not reach production in time.

Albatros D.Va Specifications

Engine:	160/170 hp Mercedes D.III/D.IIIa	
Wing:	Span	9.04 m
	Area	21.2 m^2
General:	Length	7.33 m
	Height	2.70 m
	Empty Weight	687 kg
	Loaded Weight	937 kg
Maximum Speed		185 km/h
Climb:	1000m	4.3 min
	2000m	8.8 min
	3000m	14.5 min
	4000m	22.7 min
	5000m	35.0 min

Pfalz D.IIIa Specifications

Engine:	160/170 hp Mercedes D.III/D.IIIa	
Wing:	Span Upper	9.40 m
	Span Lower	7.80 m
	Chord Upper	1.65 m
	Chord Lower	1.20 m
	Dihedral Upper	0
	Dihedral Lower	1 deg
	Gap	1.40 m
	Area	22.76 m^2
General:	Length	6.95 m
	Height	2.67 m
	Empty Weight	689 kg
	Loaded Weight	922 kg
Maximum Speed		180 km/h
Climb:	1000m	3.3 min
	2000m	7.3 min
	3000m	11.7 min
	4000m	17.3 min
	5000m	24.7 min

Above: The Albatros D.Va was the main frontline fighter in January 1918, and four of them were at the competition for comparison with the new fighters being evaluated. Two had the standard Mercedes D.III, one a high-compression Mercedes D.IIIaü, and one the new BMW.IIIa. Above is an Albatros D.Va flown by *Jasta* Boelcke.

Below: The Pfalz D.IIIa was the second most numerous frontline fighter in January 1918, and two of them were at the competition for comparison with the new fighters being evaluated. One was D.IIIa 6033/17 shown below. Both had the standard production Mercedes D.III engine.

German Fighter Aircraft Competing at the First Fighter Competition (D Flugzeug Wettbewerb) Held at Adlershof 20 January–12 February 1918

Aircraft	Engine	Remarks
A.E.G. D.I ‡	Mercedes D.III	
Albatros D.Va 7117/17	B.M.W.IIIa	Operational airframe with new engine
Albatros D.Va (factory #4563)	Mercedes D.IIIaü	Operational airframe with high comp. engine
Albatros D.Va 7089/17	Mercedes D.III	Standard operational fighter for comparison
Albatros D.Va 7090/17	Mercedes D.III	Standard operational fighter for comparison
Aviatik D.III	Benz Bz.IIIbo V-8	
DFW T 34-I	Mercedes D.III	Biplane; rejected before flight evaluations
DFW T 34-II	Mercedes D.III	The only *new* triplane design competing
Fokker V.9 ‡	Oberursel Ur.II	
Fokker V.11	Mercedes D.III	Won the competition, became Fokker D.VII
Fokker V.13-I	Oberursel Ur.III	Prototype for Fokker D.VI
Fokker V.13-II	Siemens-Halske Sh.III	
Fokker V.17	Oberursel Ur.II	Monoplane, fastest fighter at the competition
Fokker V.18	Mercedes D.III	
Fokker V.20 ‡	Mercedes D.III	Monoplane with wood cantilever wing
Fokker Dr.I 201/17	Goebel Goe.III	Operational airframe with more powerful eng.
Fokker Dr.I 469/17	Oberursel Ur.III	Operational airframe with more powerful eng.
Junkers J7	Mercedes D.III	All-metal, low-wing cantilever monoplane
Kondor D.II ‡	Oberursel Ur.II	
L.V.G. D.IV	Benz Bz.IIIbo V-8	
Pfalz D.IIIa 5935/17	Mercedes D.III	Standard operational fighter for comparison
Pfalz D.IIIa 6033/17	Mercedes D.III	Standard operational fighter for comparison
Pfalz D.VI	Oberursel Ur.II	
Pfalz D.VII	Siemens-Halske Sh.III	Evaluation batch of 50 ordered
Pfalz D.VIII ‡	Siemens-Halske Sh.III	Disqualified by persistent engine problems
Roland D.VI	Benz Bz.IIIa	Prototype of Roland D.VIb
Roland D.VI	Mercedes D.III	Evaluation batch of 50 ordered as D.VIa
Roland D.VII ‡	Benz Bz.IIIbo V-8	Disqualified by persistent engine problems
Roland D.IX	Siemens-Halske Sh.III	Crashed during competition
Rumpler D (7D4 U-strut)	Mercedes D.III	Evaluation batch of 50 ordered
Rumpler D (7D4 parallel-strut)	Mercedes D.III	
Schütte-Lanz D.III	Mercedes D.III	
Siemens-Schuckert D.III 8340/17	Siemens-Halske Sh.III	Evaluation batch of 50 ordered

Notes:

1. An ‡ indicates not listed in the official flight result tabulations, but present without competing. This could be because the possibility of production had already been discarded (e.g., AEG D.I) or the engine gave too many problems and disqualified the aircraft from official competition (e.g., Pfalz D.VIII).
2. The Fokker V11 won the competition and became the prototype for the production Fokker D.VII.
3. A batch of 50 Roland D.VI fighters was ordered for operational evaluation at the front.
4. A batch of 120 (later reduced to 60) Fokker D.VI fighters, the production version of the Fokker V.13, were ordered for operational evaluation at the front.
5. Both the Pfalz D.VII and SSW D.III powered by the 160 hp Siemens-Halske Sh.III demonstrated excellent climb and were ordered for operational evaluation at the front.
6. The Albatros D.IX was planned to compete but crashed just before the competition.
7. Because Pfalz was a Bavarian company and Bavaria wanted to maintain as much autonomy as possible, the Pfalz D.XII replaced the Pfalz D.IIIa in production despite not being at the competition.
8. Examples of the main production fighters then at the front, the Albatros D.Va, Pfalz D.IIIa, and Fokker Dr.I, were entered in the competition to provide a standard for comparison with the new designs. Two Albatros D.Va fighters and both Pfalz D.IIIa fighters had their standard engines. In addition, two more Albatros D.Va fighters had new, high-compression engines. Both Fokker Triplanes at the competition used a more powerful engine than the standard production model.

Front-Line Fighters, Winter 1917–18

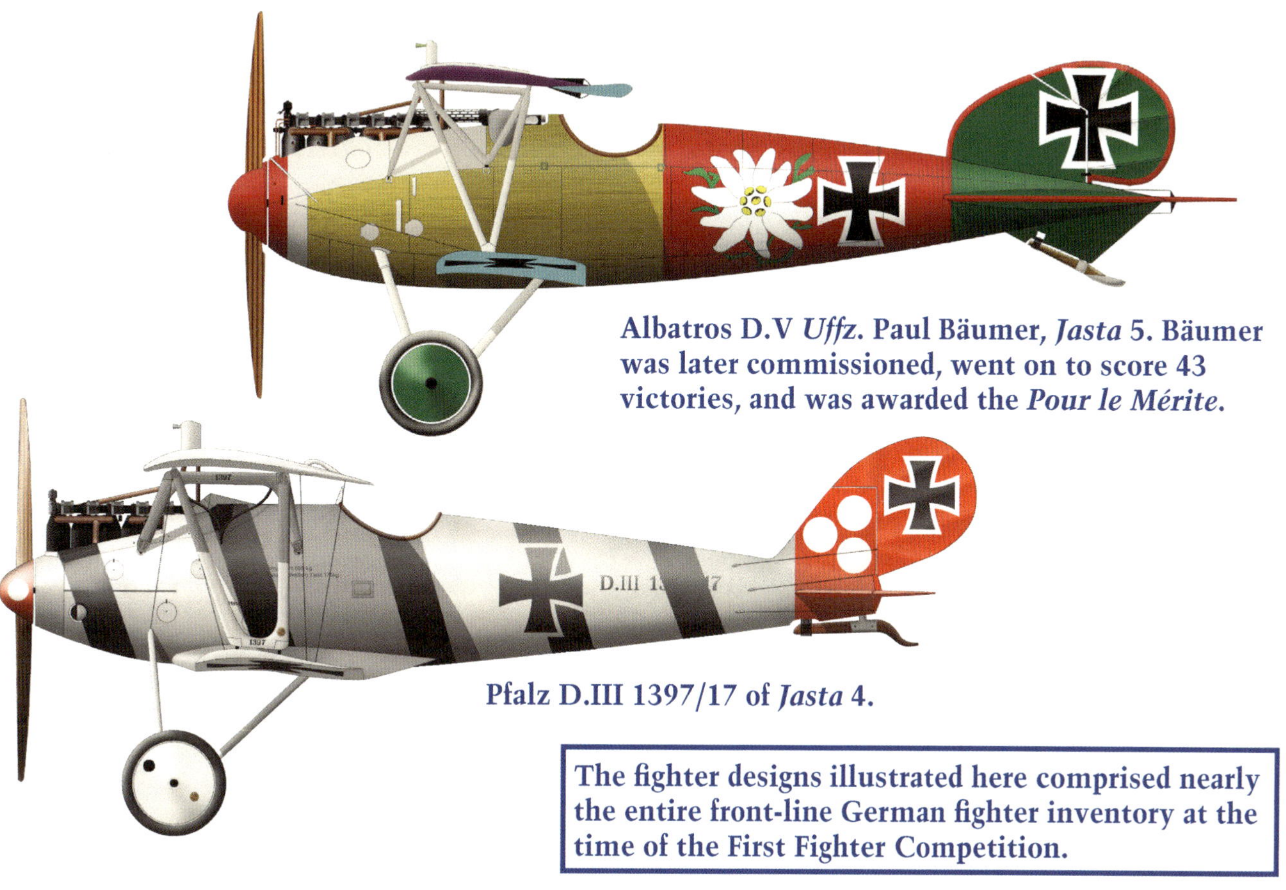

Albatros D.V *Uffz.* Paul Bäumer, *Jasta* 5. Bäumer was later commissioned, went on to score 43 victories, and was awarded the *Pour le Mérite*.

Pfalz D.III 1397/17 of *Jasta* 4.

The fighter designs illustrated here comprised nearly the entire front-line German fighter inventory at the time of the First Fighter Competition.

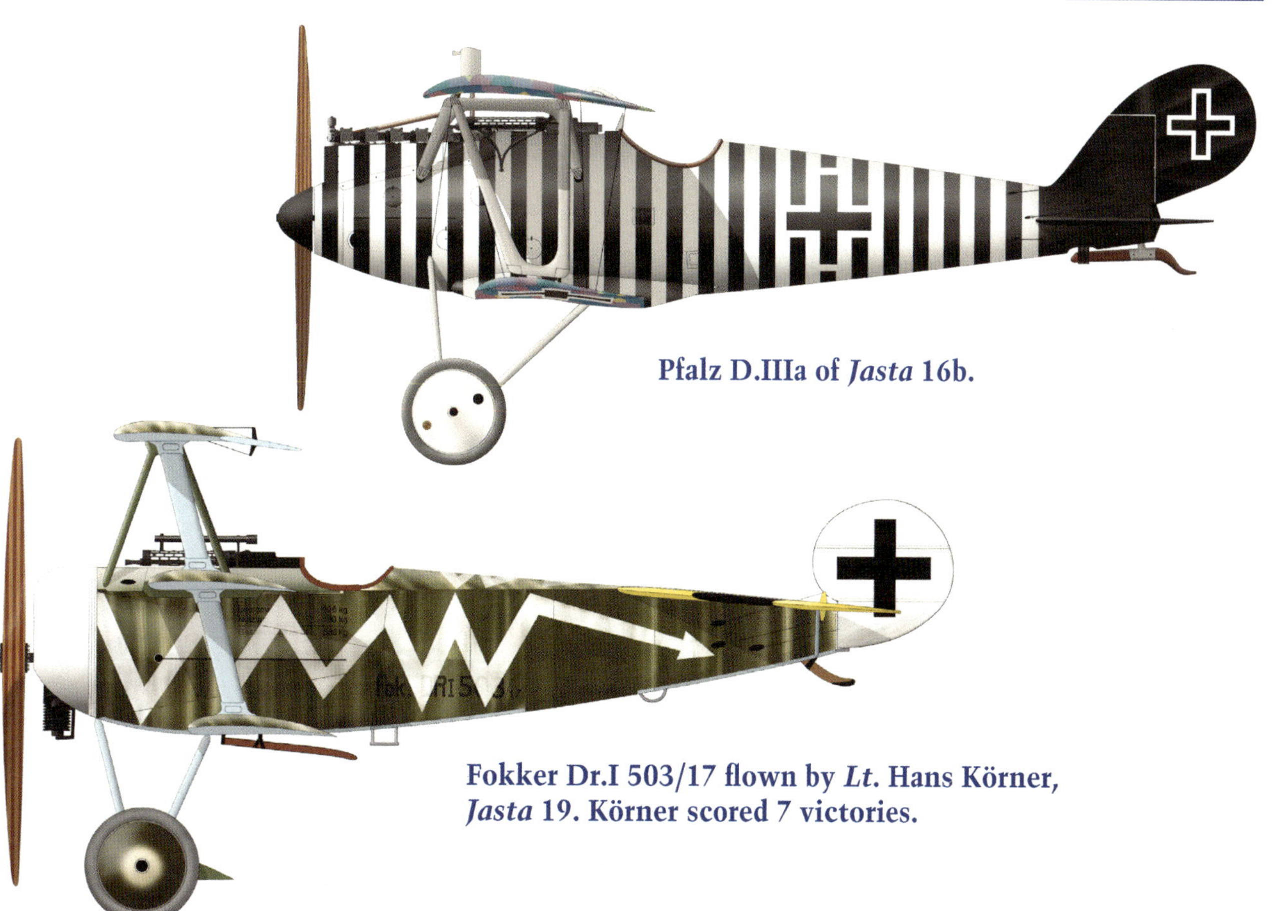

Pfalz D.IIIa of *Jasta* 16b.

Fokker Dr.I 503/17 flown by *Lt.* Hans Körner, *Jasta* 19. Körner scored 7 victories.

Fokker at the First Fighter Competition

Fokker brought the greatest variety of prototype fighters to the First Fighter Competition. In addition to Fokker Dr.I triplanes with prototype rotary engines more powerful than the production aircraft, there were rotary-powered biplanes and a mid-wing monoplane. These were joined by two similar biplanes powered by the Mercedes D.IIIa.

The rotary-powered biplane prototypes, the V9 and two versions of the V13, were biplane developments of the Fokker Triplane. The V9 used the same 110 hp Oberursel Ur.II rotary engine that was standard in the Fokker Triplane. The V9 was lighter than the Triplane and had somewhat better climb. Despite having more power, the Fokker V13-I fitted with a prototype 145 hp Oberursel Ur.III engine had slightly lower climb rate, perhaps because it was heavier. The V9 and V13 were similar, but the V13 had N-struts. The best features of these prototypes were combined to create the production Fokker D.VI that was built in small numbers. The D.VI had excellent maneuverability and good flying qualities, and despite being powered by the 110 hp Oberursel Ur.II was fast at low altitude. However, the D.VI needed more power, especially at altitude, and the more powerful rotary engines in development were not ready for mass production. This limited D.VI production in favor of the D.VII powered by the 160 hp Mercedes D.IIIa.

Fokker V9 Specifications

Engine:	110 hp Oberursel Ur.II	
Wing:	Span	7.70 m
	Area	16.5 m^2
General:	Length	5.90 m
	Height	2.80 m
	Empty Weight	381 kg
	Loaded Weight	572 kg
Climb:	1000m	2.5 min
	2000m	4.7 min
	3000m	7.7 min
	4000m	11.0 min
	5000m	15.5 min

Fokker Dr.I Specifications

Engine:	110 hp Oberursel Ur.II	
Wing:	Span Upper	7.190 m
	Span Middle	6.225 m
	Span Lower	5.725 m
	Wing Area	18.66 m^2
General:	Length	5.77 m
	Height	2.95 m
	Empty Weight	406 kg
	Loaded Weight	586 kg
Maximum Speed:		185 km/h
Climb:	1000m	2.9 min
	2000m	5.5 min
	3000m	9.3 min
	4000m	13.9 min
	5000m	21.9 min

Below: The Fokker V9 prototype used an 110 hp Oberursel U.II engine, the same engine used in the Fokker Triplane. The interplane struts differed from those in the production D.VI fighter developed from the V9 and V13.

Fokker V9, V13, & D.VI

Left & Below: The Fokker V9 was powered by a 110 hp Oberursel Ur.II. It was photographed at the Fokker factory before the First Fighter Competition. The production Fokker D.VI used the 110 hp Oberursel Ur.II like the V9 and the N-shaped interplane struts featured by the V13 prototypes.

Above: The Fokker V9, work number 1831, photographed at the First Fighter Competition.

Above: Fokker V13-II, Fokker Works number 2054, the second of two V13 prototypes, participated in the First Fighter Competition powered by a 160 hp Siemens-Halske Sh.III. The V13-I, works number 1980, was powered by the 145 hp Oberursel Ur.III. The Fokker V13 prototypes featured the N-strut design used by the production Fokker D.VI.

Below: The Fokker D.VI was a smaller, lighter sibling of the D.VII powered by a 110 hp Oberursel Ur.II. This one is serving with *Jasta* 80b. Somewhat more maneuverable than the larger D.VII and slightly faster at low altitude, the low-powered rotary limited the D.VI to production of only 60 aircraft. The more powerful inline engines in the D.VII gave it better performance at higher altitudes, where most combats started by 1918.

Right: Fokker D.VI of *Oblt*. Max Speidel of *Jasta* 80b. The center fuselage was painted black (or, because it was a Bavarian unit, possibly dark blue) with white stripes, and white stripes were added on the tailplane and between the crosses on the top of the upper wing.

Below: Fokker D.VI 1688/18, the next to last production aircraft.

Below: D.VI 1689/18, the last production aircraft, photographed 21 June 1918. The D.VI had excellent visibility for the pilot, was very maneuverable, and was light and sensitive on the controls, a real pilot's airplane. It just needed more power than its 110 hp Oberursel Ur.II provided to be a first-class fighter. Compare the climb times of the V13-I and D.VI on page 30 to see the importance of more power.

Left: Although 120 Fokker D.VI fighters were ordered, only 60 were built. The rest of the order was apparently changed to build Fokker E.V parasol monoplane fighters instead of the D.VI biplane. Both aircraft used the same 120 hp Oberursel Ur.II rotary engines, the same engine used in the Fokker Triplane. At left is a lineup of Fokker D.VI fighters at *Jasta* 80b in August 1918 at Morsberg.

Below & Below Right: *Lt.* Kurt Seit and his Fokker D.VI *Fratz* at *Jasta* 80b. *Fratz* had plenty of teeth, probably related to the fact that Seit became a dentist after the war. Bavarian colors (blue & white) were used for the personal markings.

Fokker D.VI

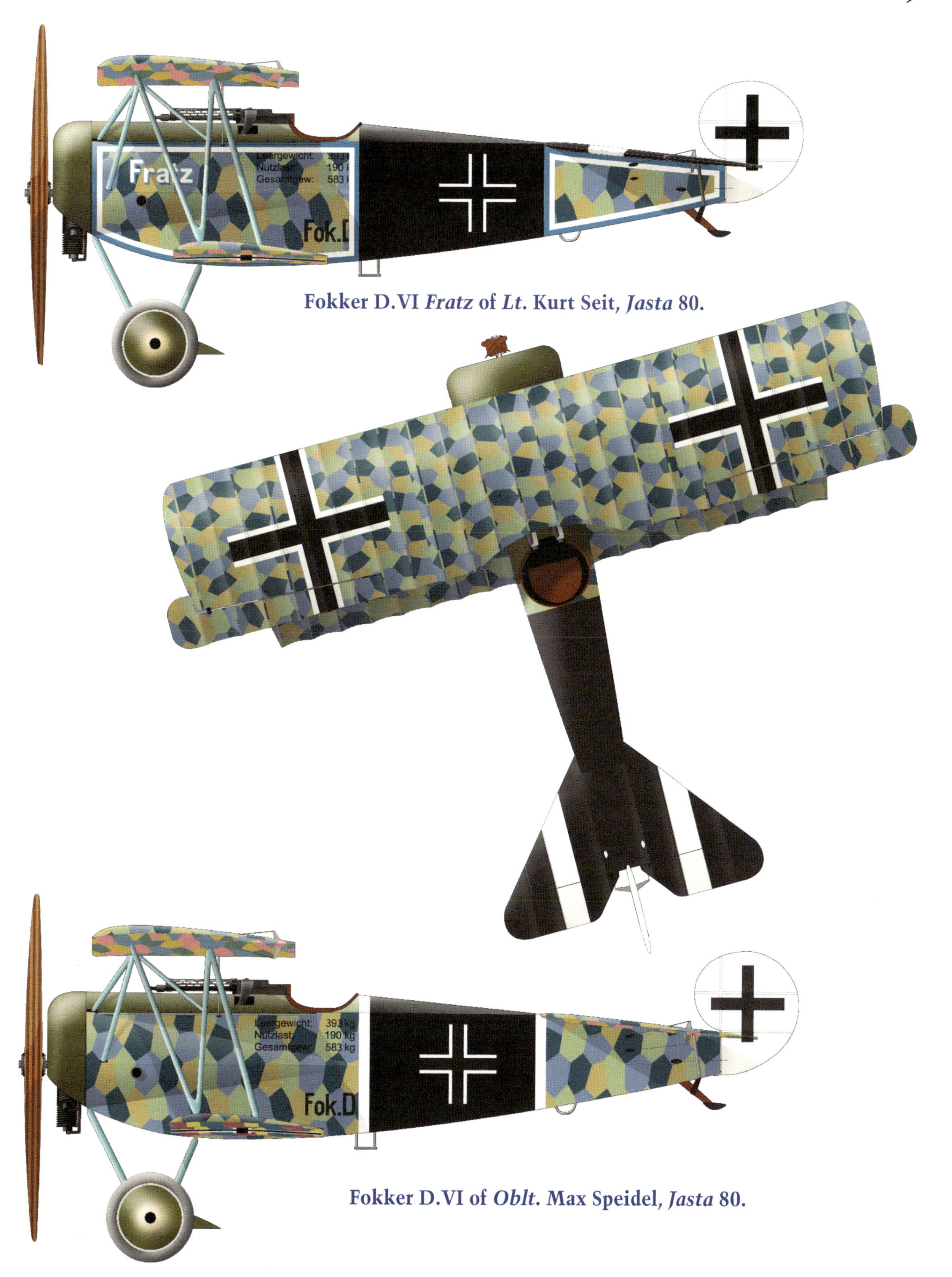

Fokker D.VI *Fratz* of *Lt.* Kurt Seit, *Jasta* 80.

Fokker D.VI of *Oblt.* Max Speidel, *Jasta* 80.

Fokker V13-I Specifications		
Engine:	145 hp Oberursel Ur.III	
Wing:	Span Upper	7.65 m
	Span Lower	5.80 m
	Area	17.7 m²
General:	Length	6.23 m
	Height	2.555 m
	Empty Weight	440 kg
	Loaded Weight	640 kg
Climb:	2000m	4.5 min
	3000m	7.7 min
	4000m	11.0 min
	5000m	16.2 min

Fokker D.VI Specifications		
Engine:	110 hp Oberursel Ur.II	
Wing:	Span Upper	7.20 m
	Span Lower	5.81 m
	Area	17.1 m²
General:	Length	5.78 m
	Height	2.65 m
	Empty Weight	395 kg
	Loaded Weight	588 kg
Maximum Speed:		196 km/h
Climb:	1000m	2.6 min
	2000m	5.9 min
	3000m	10.7 min
	4000m	17.3 min
	5000m	28.0 min

Fokker V17

Fokker's first cantilever monoplane fighter, the V17, was the final rotary-powered Fokker entry in the First Fighter Competition. The wing was built around two wood box spars covered in plywood, giving it a smooth finish and good torsional stiffness for effective aileron control. Powered by the same 110 hp Oberursel Ur.II engine used in the Fokker Triplane, it was very maneuverable and fast for its power, reportedly being the fastest fighter at the First Fighter Competition. It also had an excellent climb rate; however, its wing obscured the pilot's field of view downward and it was officially rejected for that reason. Another reason may have been that monoplanes, especially cantilever monoplanes, were viewed as structurally suspect.

Fokker V17 Specifications		
Engine:	110 hp Oberursel Ur.II	
Wing:	Span	8.375 m
	Area	11.6 m²
General:	Length	5.80 m
	Height	2.80 m
	Empty Weight	356 kg
	Loaded Weight	535 kg
Climb:	1000m	3.2 min
	2000m	6 min
	3000m	9 min
	4000m	13 min
	5000m	19 min

Below: The Fokker V17 participated in the First Fighter Competition powered by a 110 hp Oberursel Ur.II.

Above & Below: The Fokker V17, work number 2147, powered by a 110 hp Oberursel Ur.II participated in the First Fighter Competition. Despite its low power it was the fastest fighter at the competition. It also had a good climb rate and was very maneuverable. The V17 was the first Fokker cantilever monoplane design. The wood wings were built around two box spars covered with plywood, providing strength and torsional stiffness for effective aileron control. The wing obscured the pilot's downward view and monoplanes were structurally suspect, so the V17 remained a prototype.

Fokker V11

By 1918 most fighter combats started at high altitude, so a high ceiling and good speed and climb at high altitude were essential. The low-power rotary engines available in production quantities in Germany did not offer the necessary performance at high altitude. More powerful rotaries were in development but not yet ready for mass production. Of these prototype rotary engines, only the Siemens-Halske Sh.III would eventually become available in production quantities, and that only after prolonged development characterized by short engine life and frequent failures. The rotary engine problems were greatly aggravated by the almost complete lack of suitable lubricants in Germany. All these factors favored fighters with more powerful inline engines like the Mercedes D.III series.

Consequently, the Fokker V11 prototype powered by a 160 hp Mercdes D.III was greeted with greater enthusiasm at the fighter competition than its rotary-powered siblings. The V11 was based on the same technology at the rotary-powered V9 and V13, although its wing cellule with N-struts more closely resembled that of the V13. The V11 was heavier than the V9 and needed the stronger, two-spar design for its larger lower wing.

In its original form the V11 had a short fuselage without fixed vertical fin, clearly derived from its Triplane ancestor. The short fuselage without fin did not account for the additional side area of the longer nose, and the V11 was directionally unstable and dangerous to fly. It was soon modified with a longer fuselage, improving its flying characteristics. However, it was still not stable enough.

The V11 and its companion V18 were strong, structurally sound yet simple airplanes that, with some tuning of their stability, offered good performance and flying qualities at high altitude. The greater power at altitude they needed was on the way with the new, over-compressed BMW.IIIa engine, and they were the prototypes of the Fokker D.VII, the best overall fighter of the war.

Fokker V11-II Specifications

Engine:	160 hp Mercedes D.III	
Wing:	Span	8.72 m
	Area	24.0 m^2
General:	Length	6.90 m
	Height	3.05 m
	Empty Weight	592 kg
	Loaded Weight	821 kg
Climb:	1000m	3 min
	2000m	6.5 min
	3000m	11 min
	4000m	17.3 min
	5000m	22.5 min

Below: The Fokker V11, works number 1883, participated in the First Fighter Competition powered by a 160 hp Mercedes D.III engine. It is seen here in its original form with short fuselage and no fixed fin, clear indication of its Triplane heritage.

Above: The Fokker V11, work number 1883, in revised form with longer fuselage that imparted greater stability.

Below: Fokker V11 in its original form. There is slight dihedral on the upper wing and the undercarriage axle does not have an airfoil. The nose radiator was simple and effective. The wing design was excellent both structurally and aerodynamically, being strong, relatively simple, and offering exceptional stall characteristics.

Fokker V18

The Fokker V18 prototype was closely related to the V11 but had a more powerful version of the Mercede engine and, more importantly for its flying qualities, it had much enlarged vertical tail surfaces for better stability. As such, the V18 was an important step toward the Fokker D.VII production fighter that soon followed.

Below & Bottom: The Fokker V18 was based on the same formula as the V11 but had larger vertical tail surfaces for improved stability.

Fokker V18 Specifications		
Engine:	180 hp Mercedes D.IIIaü	
Wing:	Span	8.72 m
	Area	23.1 m^2
General:	Length	7.03 m
	Height	2.9 m
	Empty Weight	634 kg
	Loaded Weight	860 kg
Climb:	2000m	5.3 min
	3000m	10.5 min
	4000m	17.3 min
	5000m	26.0 min

Above: The Fokker V18, work number 2116, looks like the earlier V11 in this front view.

Fokker V20

Fokker's second cantilever monoplane fighter, the V20, was the inline engine counterpart to the V17. Construction was similar, but power was by the 160 hp Mercedes D.III for better high-altitude performance. Like the V17, its wing obscured the pilot's field of view downward, and monoplanes, especially cantilever monoplanes, were viewed as structurally suspect.

Fokker V20 Specifications

Engine:	160 hp Mercedes D.III	
Wing:	Span	9.29 m
	Area	15.2 m^2
General:	Length	7.135 m
	Height	3.0 m
	Empty Weight	620 kg
	Loaded Weight	821 kg
Climb:	2000m	6.7 min
	3000m	11.7 min
	4000m	18.0 min
	5000m	26.7 min

Left & Below: Fokker V20 monoplane prototype, work number 2219, was the inline-powered counterpart to the V17 rotary-powered prototype. The V18 offered better field of view for the pilot and time was not yet ripe for authorities to trust a cantilever monoplane design to withstand the harsh stresses of aerial combat.

Fokker's Triumph – The Fokker D.VII

Fokker had struggled since the eclipse of his *Eindeckers,* and his famous Triplane was only a partial success. Limited to the same engines and weapons his competitors used, Fokker needed a significant breakthrough to built a dramatically better airplane, and he finally achieved it in the Fokker D.VII with its wing of innovative structural and aerodynamic design.

Above: One of the greatest services Manfred von Richthofen did for Germany was instigating the fighter competitions, and the Fokker D.VII was the greatest result of those competitions. Arriving at the front days after Richthofen's death in his Fokker Triplane, he was not able to fly the D.VII in combat himself. Initially using the Mercedes D.IIIa engine, when fitted with the superb 185 hp BMW.IIIa it was the best all-around fighter of the war. This D.VII flew with *Jasta* 49.

Although sharing its engine and armament with the Albatros, Pfalz, and other German designs, the Fokker D.VII introduced important structural and aerodynamic innovations that greatly improved its effectiveness. By far the most important was its thick wooden wing built around a box spar. The thick wing, with its rounded leading edge, offered high lift and exceptional stalling characteristics, making the D.VII maneuverable and easy to fly and enabling it to 'hang on its prop' without stalling. These exceptional handling qualities made good pilots out of average ones and made aces out of good pilots. Other designers used thin airfoils because they had somewhat less drag than the thick Fokker wing. However, the strong Fokker wing eliminated the need for the extensive system of bracing wires that thin airfoils required. The combined drag of thin airfoils with their bracing wires was significantly more than the drag of the thick Fokker wing which needed no bracing wires, and this was the secret to the Fokker's improved speed and climb with the same power.

Due to stagnant engine development, upon its introduction to service the Fokker D.VII was powered by the same basic Mercedes D.IIIa engine used in the Albatros scouts in August 1916! It was not until the new BMW engine finally arrived in June/July that the Fokker D.VII fulfilled its full potential and became the premier fighter of the war. The BMW engine was similar to the familiar 160 hp Mercedes D.III engine but developed its 185 hp at 2,000 meters altitude because it was over-compressed. That meant it could not be run at full throttle until reaching the thinner air at 2,000 meters without detonation and engine damage. This design gave it more power at high altitude for increased speed and exceptional climb. Mercedes countered with the over-compressed 180 hp D.IIIaü and finally the 220 hp D.IIIavü, the latter having greater bore. Both engines were used in the D.VII, with the D.IIIavü giving performance equivalent to the BMW. Although the BMW-powered Fokker D.VII became a legend in its own time, there were never enough of them to win the air war for Germany.

Above: Reproduction Fokker D.VII displayed in authentic camouflage and markings at the USAF Museum. This D.VII is in the colors of *Lt.* Rudolf Stark, *Jastaführer* of Jasta 35b ('b' indicating a Bavarian unit), 11 victory-ace, and author of the postwar book ***Wings of War***.

Below: Reproduction D.VII in authentic camouflage and representative markings evokes the first war in the air.

Fokker D.VII Specifications

Engine:	185 hp BMW.IIIa	
Wing:	Span	8.90 m
	Area	20.2 m^2
General:	Length	6.954 m
	Height	2.945 m
	Empty Weight	688 kg
	Loaded Weight	906 kg
Maximum Speed:		200 km/h
Climb:	1000m	1.8 min
	2000m	4.0 min
	3000m	7.0 min
	4000m	10.2 min
	5000m	14.0 min
	6000m	18.7 min

Above: Fokker D.VII flown by *Lt.* Josef Mai, *Jasta* 6, September 1918. Mai scored 30 victories and was recommended for the *Pour le Mérite*, but the Kaiser abdicated before the award was approved.

Below: Pilots of *Jasta* 4 in front of a Fokker D.VII of their unit, 5 September 1918. The Fokker D.VII became the main German fighter as soon as its production could re-equip the *Jastas*.

Fokker D.VII

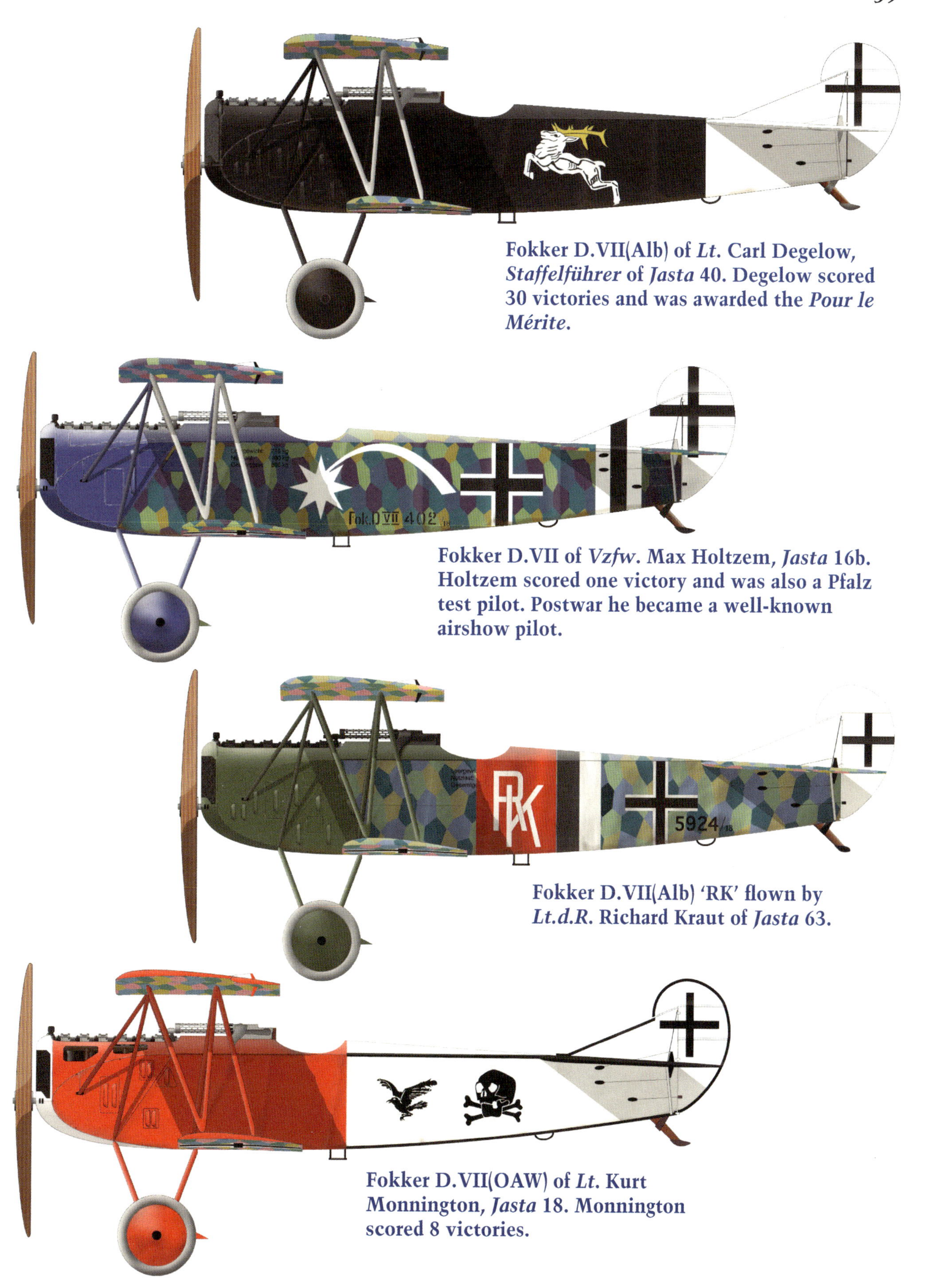

Fokker D.VII(Alb) of *Lt*. Carl Degelow, *Staffelführer* of *Jasta* 40. Degelow scored 30 victories and was awarded the *Pour le Mérite*.

Fokker D.VII of *Vzfw*. Max Holtzem, *Jasta* 16b. Holtzem scored one victory and was also a Pfalz test pilot. Postwar he became a well-known airshow pilot.

Fokker D.VII(Alb) 'RK' flown by *Lt.d.R.* Richard Kraut of *Jasta* 63.

Fokker D.VII(OAW) of *Lt*. Kurt Monnington, *Jasta* 18. Monnington scored 8 victories.

Above: An early-production Fokker D.VII flown by *Lt.d.R.* Richard Wenzel, *Jasta* 6, June 1918. Wenzel scored 12 victories and was acting commander of *Jasta* 6 for a month. His Fokker has had additional cooling Louvers punched into its upper cowling. In the heat of summer a few early D.VII fighters experienced their ammunition catching fire due to excessive heat.

Below: Fokker D.VII fighters of *JG*II in the Summer of 1918. *JG*II aircraft all featured blue fuselages and each *Jasta* had a different nose color; *Jasta* 12 had white noses, *Jasta* 13 had green noses, *Jasta* 15 had red noses, and *Jasta* 19 had yellow noses. Each pilot then added his own individual markings. See the facing page for examples.

Fokker D.VII Fighters of *JG*II

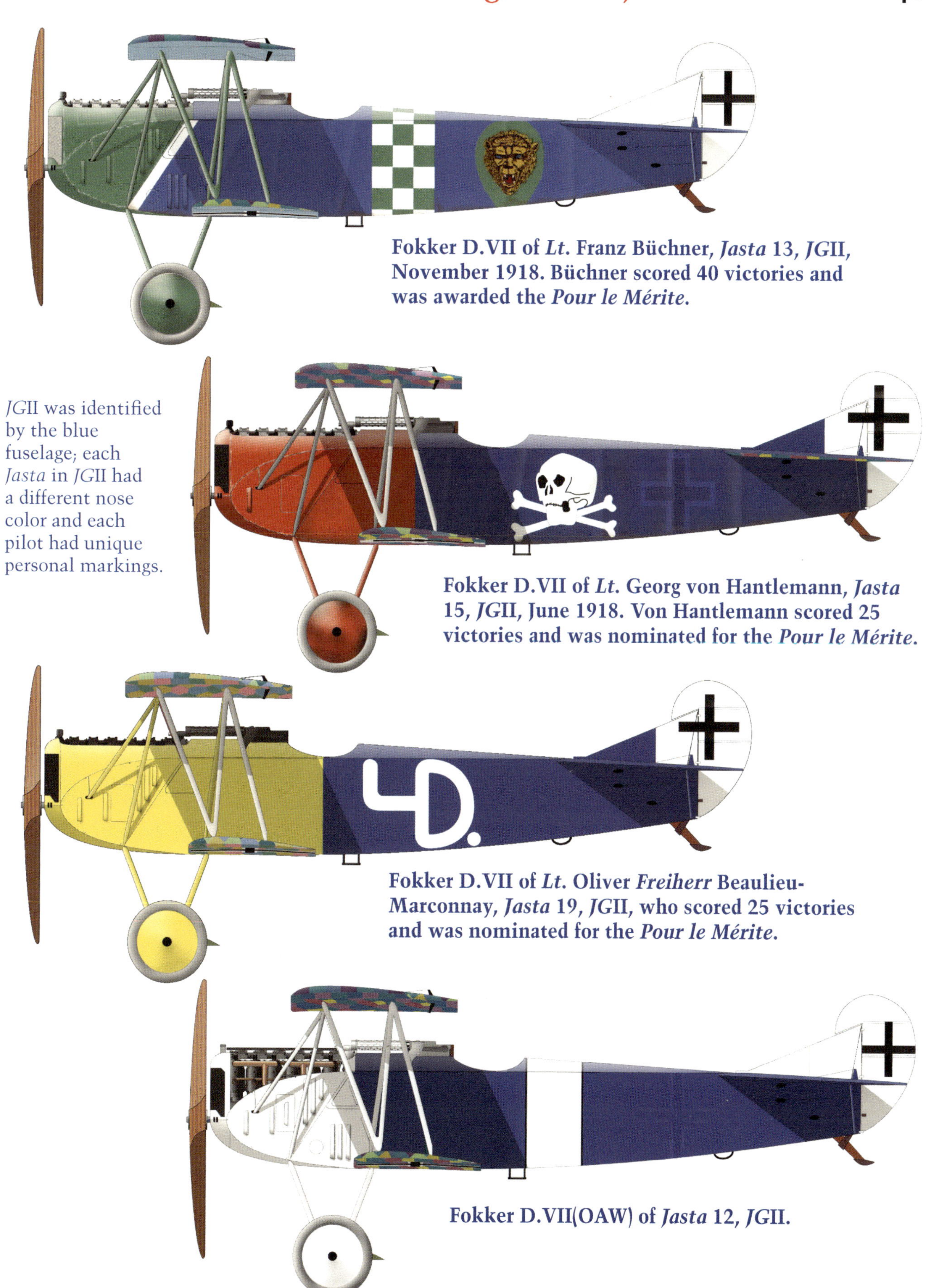

Fokker D.VII of *Lt.* Franz Büchner, *Jasta* 13, *JG*II, November 1918. Büchner scored 40 victories and was awarded the *Pour le Mérite.*

*JG*II was identified by the blue fuselage; each *Jasta* in *JG*II had a different nose color and each pilot had unique personal markings.

Fokker D.VII of *Lt.* Georg von Hantlemann, *Jasta* 15, *JG*II, June 1918. Von Hantlemann scored 25 victories and was nominated for the *Pour le Mérite.*

Fokker D.VII of *Lt.* Oliver *Freiherr* Beaulieu-Marconnay, *Jasta* 19, *JG*II, who scored 25 victories and was nominated for the *Pour le Mérite.*

Fokker D.VII(OAW) of *Jasta* 12, *JG*II.

Roland at the First Fighter Competition

Above: The Roland D.VI prototype was the 1,000th Roland-built aircraft. Powered by the proven 160 hp Mercedes D.III, it was the protoype for the D.VIa. Another D.VI powered by the 185 hp Benz Bz.IIIa was the prototype for the Roland D.VIb.

Earlier Roland fighters were strong and fast but also had poor maneuverability and handling qualities. With the showing of the D.VI prototypes at the First Fighter Competition, Roland finally delivered a fighter that was maneuverable and handled well. Two aircraft powered by different engines competed, and although the Roland D.VI did not win the contest it did well enough to earn a production contract for 50 fighters for evaluation.

The Roland D.VIa with 160 hp Mercedes D.III was somewhat faster and more maneuverable than the Albatros and Pfalz fighters it was meant to replace, but the pilots were hoping for a greater advance. The more powerful Roland D.VIb offered better speed and climb and was a bigger step in the right direction.

Combat experience soon showed the superiority of the Fokker D.VII in maneuverability and handling qualities, particularly its controllability at high angles of attack and benign stall characteristics, and the Fokker became an immediate pilot favorite. The Roland D.VI was a good fighter but could not compare with the Fokker D.VII, and the D.VI was the last Roland fighter to enter service. However, limited production lasted until the war's end.

Roland D.VIa Specifications

Engine:	160 hp Mercedes D.III	
Wing:	Span Upper	9.40 m
	Span Lower	8.68 m
	Chord Upper	1.40 m
	Chord Lower	1.20 m
	Gap	1.49 m
	Area	21.7 m²
General:	Length	6.40 m
	Height	2.60 m
	Empty Weight	655 kg
	Loaded Weight	845 kg
Maximum Speed:		190 km/h
Climb:	1000m	2.5 min
	2000m	6.0 min
	3000m	11.0 min
	4000m	18.0 min
	5000m	25.0 min

Roland D.VIb Specifications

Engine:	185 hp Benz Bz.IIIa	
Wing:	Span Upper	9.40 m
	Span Lower	8.68 m
	Chord Upper	1.40 m
	Chord Lower	1.20 m
	Gap	1.49 m
	Area	21.7 m²
General:	Length	6.40 m
	Height	2.60 m
	Empty Weight	655 kg
	Loaded Weight	845 kg
Maximum Speed:		200 km/h
Climb:	1000m	2.5 min
	2000m	4.9 min
	3000m	7.8 min
	4000m	12.3 min
	5000m	19.0 min

Above: Although the Roland D.VI did not win the First Fighter Competition, it made a good enough showing that 50 were ordered for operational evaluation at the front. This spectacular example was flown by *Pour le Mérite* ace Otto Kissenberth with *Jasta* 23b.

Left & Below: The second D.VI prototype flew at the First Fighter Competition; different tail configurations were explored.

Roland D.VI

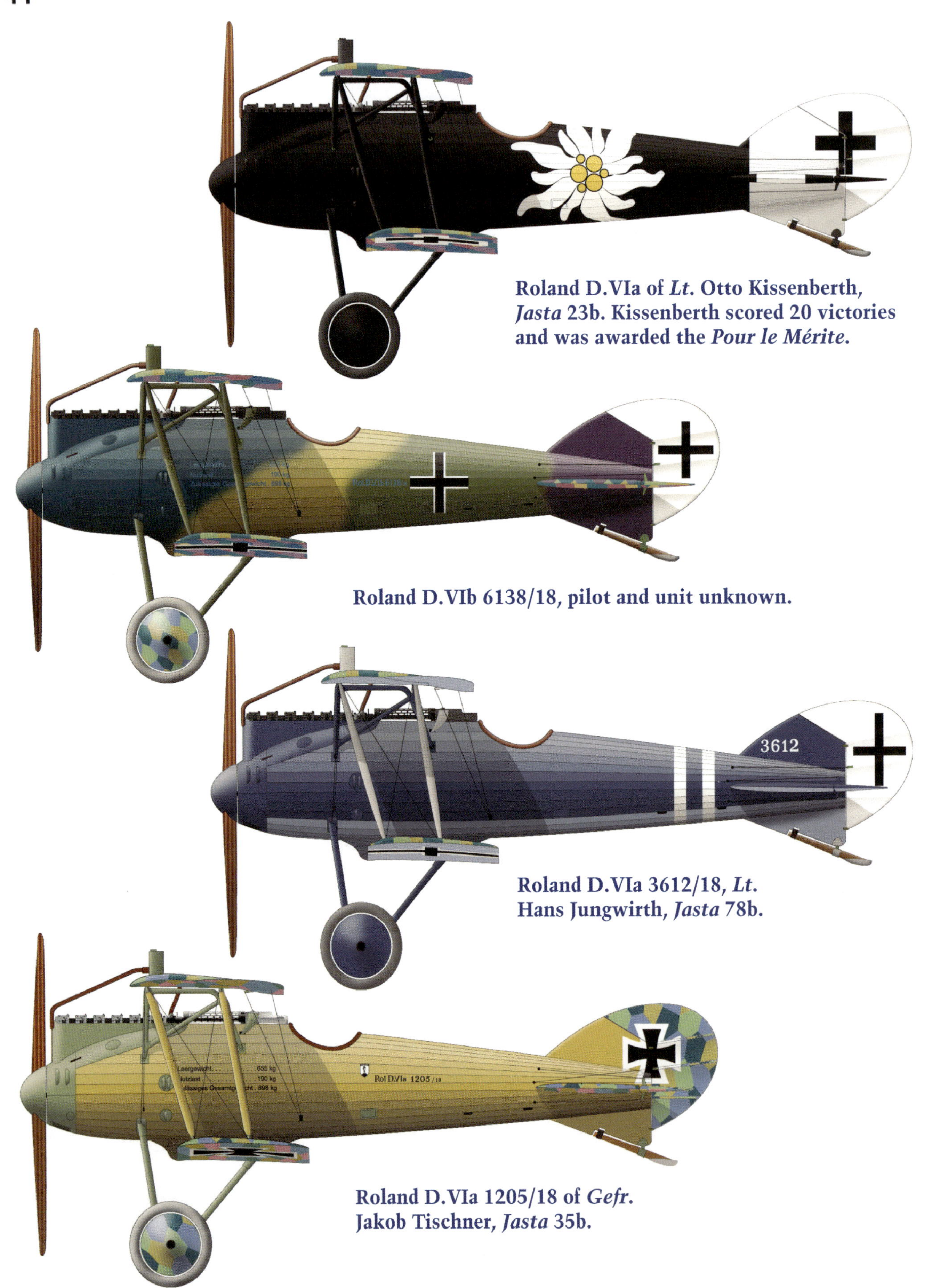

Roland D.VIa of *Lt.* Otto Kissenberth, *Jasta* 23b. Kissenberth scored 20 victories and was awarded the *Pour le Mérite.*

Roland D.VIb 6138/18, pilot and unit unknown.

Roland D.VIa 3612/18, *Lt.* Hans Jungwirth, *Jasta* 78b.

Roland D.VIa 1205/18 of *Gefr.* Jakob Tischner, *Jasta* 35b.

Roland D.VII & D.VIII

Above: The Roland D.VII prototype was powered by the experimental Benz Bz.IIIbo ungeared V-8.

The Roland D.VII prototype was developed in parallel with the Roland D.VI; the key difference was the D.VI used the production Mercedes inline six whereas the D.VII used the experimental 185 hp Benz Bz.IIIbo V-8. The more powerful V-8 gave better performance when it ran reliably, but the V-8 engine continually gave problems. After participating in the First and Second Fighter Competitions, further development was terminated due to engine problems.

The Roland D.VIII differed from the D.VII by having a geared 185 hp Benz Bz.IIIbm V-8. That engine was even more troublesome than its ungeared counterpart, and after appearing at the Second Fighter Competition, development of the Roland D.VIII was also terminated due to the engine.

Roland D.VII Specifications		
Engine:	185 hp Benz Bz.IIIbo V-8	
Wing:	Span	8.84 m
	Area	20.83 m^2
General:	Length	6.10 m
	Height	2.80 m
	Empty Weight	666 kg
	Loaded Weight	858 kg
Maximum Speed:		190 km/h
Climb:	5000m	16.2 min

Right: Different tail surfaces were tried on the Roland D.VII prototype but the aircraft was doomed by its troublesome engine. Development of the similar Roland D.VIII, powered by the geared version of this engine, was likewise terminated due to engine problems.

Roland D.IX

The appearance of more powerful rotary engine prototypes in early 1918 was the impetus for development of rotary-powered fighter prototypes using these engines. The Roland D.IX was powered by the 160 hp Siemens-Halske Sh.III counter-rotary engine and appeared at the First Fighter Competition along with similarly-powered fighter prototypes from Pfalz and SSW.

The Roland D.IX demonstrated excellent performance at the First Fighter Competition but was destroyed in a freak accident. The pilot's seat collapsed in flight and the pilot inadvertently pulled the aircraft into a tight loop. The resulting high G forces drove the unfortunate pilot through the bottom of the fuselage and the aircraft broke up.

Small numbers of the competing Pfalz D.VIII and SSW D.III fighters powered by the Siemens-Halske Sh.III were ordered, and the two additional Roland D.IX fighters being built were successfully load tested and competed at the Second Fighter Competition.

Roland D.IX Specifications

Engine:	160 hp Siemens-Halske Sh.III	
Wing:	Span	8.92 m
	Area	18.48 m^2
General:	Length	5.90 m
	Height	2.75 m
	Empty Weight	534 kg
	Loaded Weight	724 kg
Maximum Speed:		185 km/h
Climb:	5000m	16.4 min

Right: The crash of the Roland D.IX fighter prototype at the First Fighter Competition was the result of a freak accident.

Below: The Roland D.IX fighter prototype was powered by the 160 hp Siemens-Halske Sh.III counter-rotary. This first prototype with overhung ailerons is the aircraft that crashed at the First Fighter Competition.

Albatros at the First Fighter Competition

Well before the First Fighter Competition Albatros was aware that its workhorse D.Va was no longer competitive and designed the D.IX, using the same engine but with a wing cellule based on that of the Spad fighter. The D.IX was completed in late 1917, but it crashed on January 18, 1918 while being prepared for the competition. This eliminated any chance Albatros, then the largest fighter supplier, had to win the competition.

The crash investigation determined the most likely cause of the crash was wing failure, and *Idflieg* cancelled the D.IX. In any case, the D.IX had demonstrated mediocre performance.

Albatros D.IX Specifications

Engine:	170 hp Mercedes D.IIIa	
Wing:	Span Upper	10.40 m
	Span Lower	10.20 m
	Chord Upper	1.30 m
	Chord Lower	1.30 m
	Gap	1.32 m
	Area	24.0 m^2
General:	Length	6.66 m
	Height	2.75 m
	Empty Weight	416 kg
	Loaded Weight	606 kg
Maximum Speed:		165 km/h
Climb:	1000m	4.0 min
	2000m	8.8 min
	3000m	14.8 min
	4000m	22.8 min
	5000m	35.0 min

Left: Front view of the Albatros D.IX fighter prototype, thought to be 2204/18, showing its Spad-type wing cellule.

Above: The Albatros D.IX fighter prototype, powered by the 180 hp Mercedes D.IIIa, crashed while being prepared for the First Fighter Competition, eliminating any chance Albatros had of winning the competition.

Pfalz at the First Fighter Competition

Above: Powered by the innovative 160 hp Siemens-Halske Sh.III counter-rotary engine, the Pfalz D.VII looked like a biplane conversion of the earlier Pfalz Dr.I. Here is is photographed at the First Fighter Competition on 23 January 1918.

Below: The Pfalz D.VIII was a two-bay version of the D.VII also powered by the Sh.III. Here is is photographed at the First Fighter Competition on 23 January 1918; the D.VII prototype is on the left. Although present at the competition, the D.VIII was unable to compete because of problems with its pre-production engine. The extra bay of struts made the D.VIII stronger, so it was selected for production, but the additional weight and drag reduced speed and climb compared to the D.VII. By the Armistice Pfalz had evidentally been able to strengthen the D.VII wing cellule because about 30 production Pfalz D.VII fighters were found postwar by the Inter-Allied Control Commission.

The Pfalz D.IIIa was in production and front-line service at the time of the First Fighter Competition and two were included in the competition to provide a standard of comparision for the experimental fighters. In addition, Pfalz brought three rotary-engine prototypes to the competition.

First was the Pfalz D.VI powered by the 110 hp Oberursel Ur.II, the standard engine in the Fokker Triplane. The D.VI had appeared a year earlier and despite fairly good performance, flying qualities, and maneuverability, nothing had been done to place it in production when it could have been competitive with Allied fighters. By the time it appeared at the competition it was over-shadowed by its siblings the D.VII and D.VIII, both of which had the considerably more powerful 160 hp Siemens-Halske Sh.III. Clearly descended from the Pfalz Dr.I that used the same engine, the D.VIII was simply a two-bay version of the D.VII, but was unable to compete because of problems with its pre-production engine.

The Pfalz D.VII made a good impression at the competition, having an excellent climb rate due to its engine, and a small production order was given for either the D.VII or D.VIII, depending on load test results. The stronger D.VIII was chosen for production and small numbers went to the *Jastas* and *KESTs*, but production was limited by availability of its engine. The D.VIII had similar performance to the competing SSW D.III but was not as maneuverable, so tended to be used more by the *KESTs* to intercept bombers. Apparently Pfalz later modified the D.VII wing to meet the load test requirements and placed it in production because about 30 D.VII fighters were found postwar by the Inter-Allied Control Commission. The reduced weight and drag of the single-bay D.VII gave it somewhat better speed than the D.VIII.

Pfalz D.VI Specifications

Engine:	110 hp Oberursel Ur.II	
Wing:	Span Upper	7.08 m
	Span Lower	6.30 m
	Chord Upper	1.30 m
	Chord Lower	1.00 m
	Area	13.3 m²
General:	Empty Weight	416 kg
	Loaded Weight	606 kg
Climb:	5000m	16 min
	6000m	25 min

Above: The elegant Pfalz D.VI, powered by a 110 hp Oberursel Ur.II rotary, did not go into production despite its good maneuverability and flying qualities. The reason for this is not known but may have been connected with the general shortage of lubricants for rotary engines in Germany. It appeared in January 1917, when the Albatros reigned supreme, and that may have been the real reason it was not placed in production. By the time of the First Fighter Competition, more powerful engines were generally required for the necessary speed and the D.VI was overshadowed by the faster Pfalz D.VII and D.VIII powered by the 160 hp Siemens-Halske Sh.III.

Pfalz D.VII Specifications

Engine:	160 hp Siemens-Halske Sh.III	
	145 hp Oberursel Ur.III	
	200 hp Goebel Goe.III	
Data for Sh.III production version:		
Wing:	Span Upper	7.52 m
	Span Lower	6.98 m
	Chord Upper	1.30 m
	Chord Lower	1.30 m
	Gap	1.45 m
	Stagger	0.25 m
	Area	17.12 m^2
General:	Length	5.65 m
	Height	2.85 m
	Empty Weight	520 kg
	Loaded Weight	715 kg
Maximum Speed:		190 kmh
Climb:	1000m	1.8 min
	2000m	4.0 min
	3000m	6.6 min
	4000m	9.8 min
	5000m	13.8 min
	6000m	21.3 min

Pfalz D.VIII Specifications

Engine:	160 hp Siemens-Halske Sh.III	
	145 hp Oberursel Ur.III	
	200 hp Goebel Goe.III	
Data for Sh.III production version:		
Wing:	Span Upper	7.52 m
	Span Lower	6.98 m
	Chord Upper	1.30 m
	Chord Lower	1.30 m
	Gap	1.45 m
	Stagger	0.25 m
	Area	17.12 m^2
General:	Length	5.65 m
	Height	2.85 m
	Empty Weight	542 kg
	Loaded Weight	767 kg
Maximum Speed:		180 kmh
Climb:	1000m	1.5 min
	2000m	3.5 min
	3000m	5.8 min
	4000m	8.4 min
	5000m	11.1 min

Left: The single-bay Pfalz D.VII first flew in December 1917 and demonstrated an excellent climb rate. The D.VII was given a small production order based on its performance at the First Fighter Competition. In the event, the stronger two-bay D.VIII was built instead.

Right: This Pfalz D.VIII served with *Jasta* 14 along with at least two others. Because the SSW D.III was somewhat more maneuverable than the D.VIII, the SSW fighters were typically assigned to front-line *Jastas* while the Pfalz was typically assigned to interceptor units. However, some reached the *Jastas* as shown here.

Pfalz D.VI, D.VII, & D.VIII

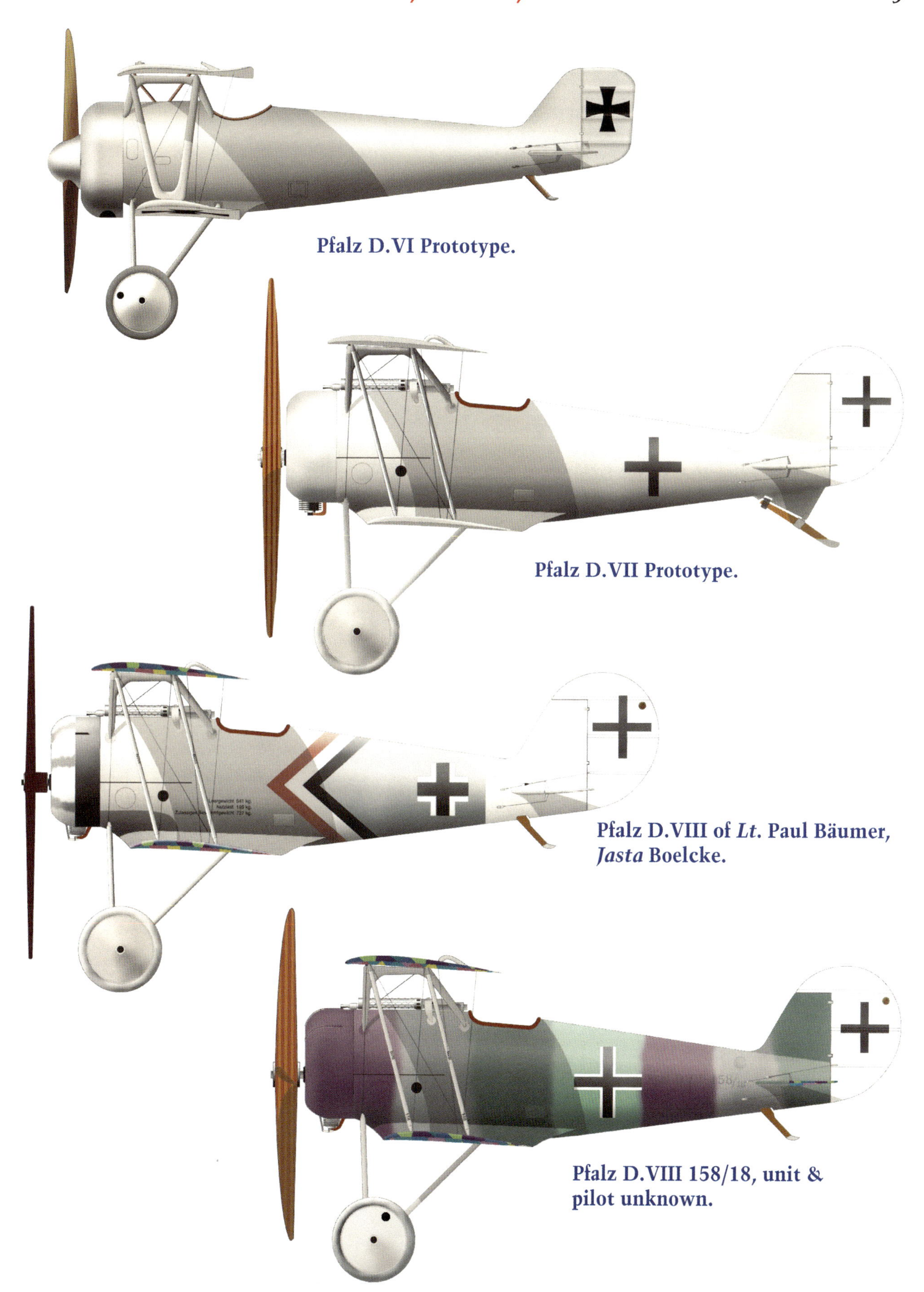

Pfalz D.VI Prototype.

Pfalz D.VII Prototype.

Pfalz D.VIII of *Lt.* Paul Bäumer, *Jasta* Boelcke.

Pfalz D.VIII 158/18, unit & pilot unknown.

SSW at the First Fighter Competition

The Siemens-Schuckert Werke entered the fighter business by building a copy of the Nieuport sesquiplane fighter. This fighter, powered by the 110 hp Siemens-Halske Sh.I nine-cylinder engine, entered limited production as the SSW D.I. The sesquiplane configuration was quickly out-moded and SSW developed a new fighter, the single-bay D.II with two-spar wings and the larger, more powerful 160 hp Siemens-Halske Sh.III eleven-cylinder engine.

Idflieg ordered 20 improved SSW D.III fighters on 26 December 1917, and several competed in the First Fighter Competition. The SSW D.III immediately demonstrated impressive climb performance but problematic maneuverability and handling characteristics. Landings, always the most delicate flight operation, were especially challenging in the D.III prototypes, partly due to the tall undercarriage required to clear the large propellers needed to absorb the power from the slow-turning counter-rotary. However, its performance was too promising to ignore, and development of both the engine and airframe continued along with introduction into small-scale service as engine availability permitted.

SSW D.III Specifications

Engine:	160 hp Siemens-Halske Sh.III	
Wing:	Span Upper	8.400 m
	Span Lower	8.130 m
	Chord Upper	1.460 m
	Chord Lower	1.000 m
	Area	18.84 m^2
General:	Length	5.850 m
	Height	2.630 m
	Empty Weight	523 kg
	Loaded Weight	725 kg
Maximum Speed:		177 km/h
Climb:	1000m	1.8 min
	2000m	3.8 min
	3000m	6.0 min
	4000m	9.0 min
	5000m	13.0 min
	6000m	20.0 min
	8100m	36.0 min

Once in service the D.III proved to have exceptional climb and maneuverability, yet needed more engine development to achieve satisfactory reliability. Its pilots thought it was the war's best fighter.

Right & Below: The SSW D.III was powered by the same 160 hp Siemens-Halske Sh.III used in the Pfalz D.VIII; the SSW D.III had similar performance but was more maneuverable than the D.VIII. These photos show an early production SSW D.III assigned to *Lt.* Walter Goettsch of *Jasta* 19. Fokker Triplanes are lined up in the background.

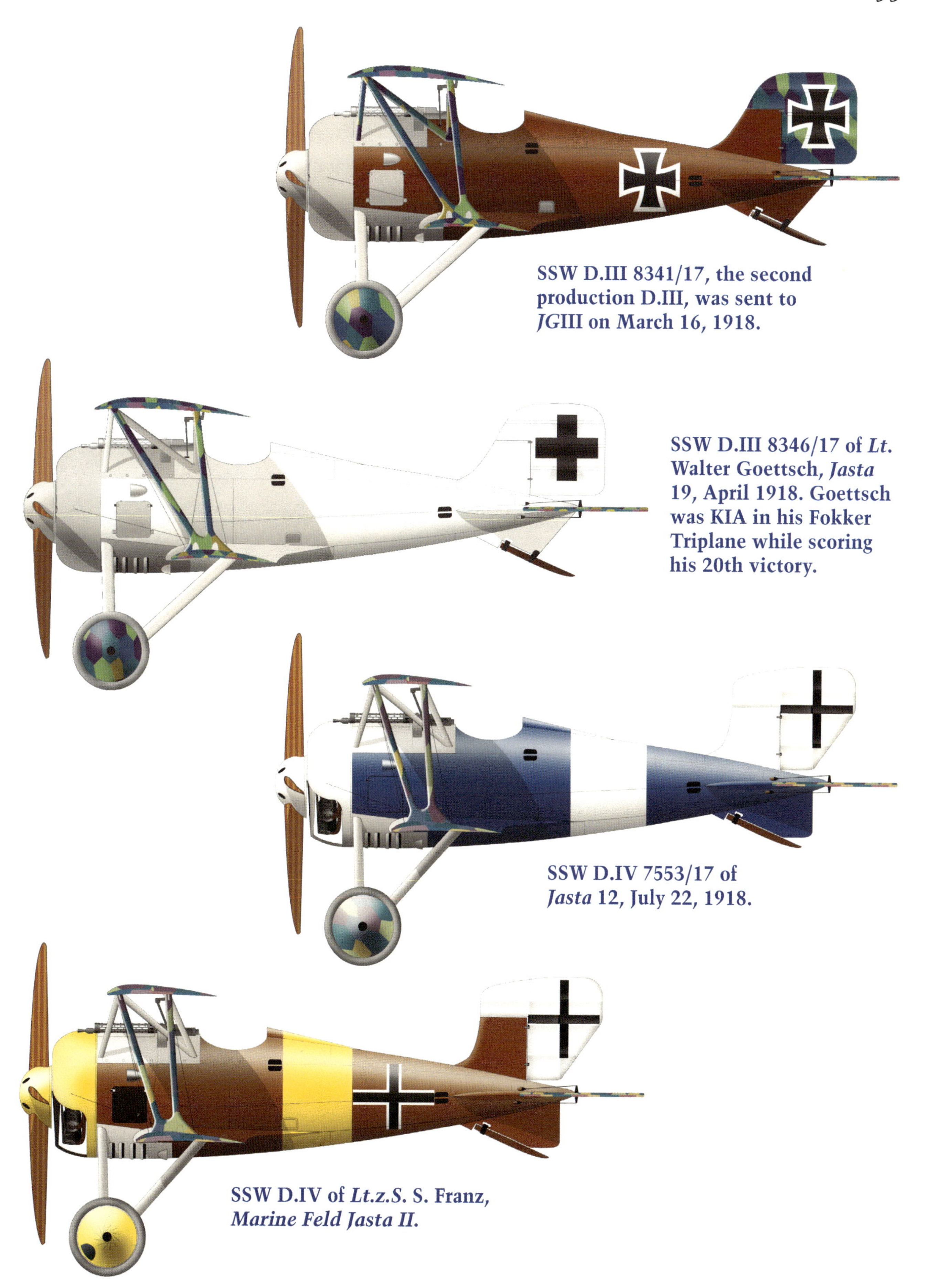

SSW D.III 8341/17, the second production D.III, was sent to *JG*III on March 16, 1918.

SSW D.III 8346/17 of *Lt.* Walter Goettsch, *Jasta* 19, April 1918. Goettsch was KIA in his Fokker Triplane while scoring his 20th victory.

SSW D.IV 7553/17 of *Jasta* 12, July 22, 1918.

SSW D.IV of *Lt.z.S.* S. Franz, *Marine Feld Jasta II.*

AEG at the First Fighter Competition

The AEG D.I was completed in May 1917. Powered by a 160 hp Mercedes D.III, it was built of steel tubing covered by fabric like all AEG aircraft.

Testing soon revealed the D.I as the fastest fighter of its day, with a speed of 225 km/h (137 mph), and its climb was equal to the new Albatros D.V that had just entered service. Its potential was obvious and *Idflieg* ordered three additional prototypes for testing. After passing its load tests on 4 August 1917, additional fight evaluation began. On 21 August *Idflieg* test pilot *Lt.* Julius Hendrichs was killed when the D.I he was flying crashed out of control. The cause was not determined, and flight evaluation continued on 25 August using another protoype.

Lt. Walter Höhndorf, *Jastaführer* of *Jasta* 14, had been involved with design of the AEG D.I. A prototype was delivered to Höhndorf at *Jasta* 14, and on 5 September Höhndorf was killed while flying D.I prototype 4400/17 when it spun out of control. With two unexplained fatal crashes, *Idflieg* cancelled planned D.I production. Given its history, the AEG D.I was not an official competitor at the First Fighter Competition and was no doubt there because it was a fast fighter with which to compare the others.

AEG D.I Specifications

Engine:	160 hp Mercedes D.III	
Wing:	Span	8.50 m
	Area	16.14 m^2
General:	Length	6.10 m
	Height	2.65 m
	Empty Weight	685 kg
	Loaded Weight	940 kg
Maximum Speed:		225 km/h
Climb:	1000m	2.2 min

Above: The AEG D.I in which Lt. Höhndorf was killed. Ear radiators were used on this prototype.

Above: The AEG D.I fighter prototype, powered by a Mercedes D.III, was present at the First Fighter Competition without competing. This may have been due to its excellent speed of 225 km/h, but by this time development had already been abandoned due to two unexplained fatal crashes. The first prototype with nose radiator is shown here.

Kondor at the First Fighter Competition

The history of the Kondor fighters is obscure. The Kondor Flugzeugwerke gmbH was founded in July 1912 to build and sell aircraft. Like many early aircraft manufacturers it built aircraft for its associated flying school and trained pilots for the air service under contract. Kondor built a number of different prototype trainers and manufactured Albatros B.II trainers under license.

Kondor built a prototype triplane fighter and a biplane derivative, the D.VII. Neither was developed beyond a single prototype, and the next Kondor fighter was the D.I prototype flown in Autumn 1917. These numbers are out of sequence and are factory dsignations, not *Idflieg* designations.

The D.I used a sesquiplane wing cellule and was powered by a 110 hp Oberursel Ur.II, the engine used in the Fokker Triplane. The wing design was obsolete and performance was disappointing, perhaps the reason for the disparagng nickname *Kondorlaus* (Kondor Louse). Despite some confusion with designations, which were factory designations, the D.I is likely is the Kondor fighter listed as present at the First Fighter Competition without officially competing. (The 'D.II' was listed as the Kondor at the First Fighter Competition, but was not completed by the time the competition was held.)

Kondor D.I Specifications

Engine:	110 hp Oberursel Ur.II	
Wing:	Span	7.60 m
	Wing Area	13.35 m²
General:	Length	4.85 m
	Height	2.40 m
	Empty Weight	388 kg
	Loaded Weight	568 kg

Kondor D.VII Specifications

Engine:	160 hp Mercedes D.III	
Wing:	Span	8.50 m
	Wing Area	15.70 m²
General:	Length	6.20 m
	Height	2.30 m
	Empty Weight	590 kg
	Loaded Weight	785 kg
Maximum Speed:		180 km/h

Left: The Kondor D.VII had a 160 hp Mercedes D.III and was a biplane derivative of an earlier Kondor triplane fighter. The gap between the wings was enlarged by suspending the lower wing below the fuselage.

Right: The Kondor D.I was powered by a 110 hp Oberursel U.II. Built in the Autumn of 1917, it had an out-moded Nieuport-type sesquiplane wing cellule with single-spar lower wing that gave so many structural problems in the Nieuport and Albatros fighters. This is the only known photograph of the type, which remained unarmed.

Aviatik at the First Fighter Competition

The Automobil und Aviatik AG of Leipzig, known as Aviatik, built the Halberstadt D.II under license as the Aviatik D.I. When *Idflieg* rationalized aircraft designations, it was logically redesignated the Halberstadt D.II(Av).

The Aviatik D.II, an original design, was built in late 1916 and demonstrated mediocre performance.

The Aviatik D.III, powered by an ungeared, experimental 195 hp Benz Bz.IIIbo V-8, flew for the first time in November 1917 and competed in the First Fighter Competition. The lower wing was mounted on a keel to give adequate gap between the

Aviatik D.II Specifications		
Engine:	160 hp Mercedes D.III	
Wing:	Span	8.84 m
General:	Length	6.82 m
	Height	2.87 m
Maximum Speed:		150 kmh
Climb:	1000m	7.2 min

Right: The Aviatik D.II was powered by a 160 hp Mercedes D.III and had mediocre performance. Only one was built in late 1916.

Below: The Aviatik D.III was powered by an ungeared, experimental 195 hp Benz Bz.IIIbo V-8. Flown for the first time in November 1917, it competed in the First Fighter Competition.

wings. The forward fuselage was of steel tube with plywood skinning and the wings were fabric covered. The type test was performed from 9–12 February 1918, at the end of the First Fighter Competition.

The Aviatik D.III did not make a big impression during the competition but a small number were built for test and evaluation. Performance was considered superior to the Albatros D.V, although that was not a particularly high standard by this time. Development of the Aviatik D.III was continued and it also competed at the Second Fighter Competition along with the Aviatik D.IV.

Aviatik D.III Specifications

Engine:	195 hp Benz Bz.IIIbo V-8	
Wing:	Span .	9.0 m
	Area	21.0 m^2
General:	Loaded Weight	864 kg
Climb:	1000m	2.5 min
	2000m	5.7 min
	3000m	11.0 min
	4000m	17.0 min

Below & Bottom: The Aviatik D.III fighter prototype was powered by a Benz Bz.IIIb high-speed V-8 engine that was not yet ready for production.

Junkers at the First Fighter Competition

Development of the all-metal Junkers J 7 single-seat fighter and its J 8 two-seat fighter sibling began in late Spring 1917. Wind tunnel tests showed that a low-mounted wing gave greater lift and the structure could absorb more of the impact of a crash, so the low wing position was chosen for the new fighters, an unusual step at the time. Even more innovative was its all-metal structure.

The J 7 initially had rotating wing-tip ailerons and a large radiator mounted over the engine to facilitate flight testing. First flight was on 18 September 1917, and despite the high-drag radiator installation, both speed and climb were promising. A new wing with conventional ailerons was then fitted and flight testing resumed on 10 October. *Lt.* Gotthard Sachsenberg and *Lt.* Theo Osterkamp, both naval aces, were able to fly the J 7 on 22 October. They were able to easily out-run and out-climb an Albatros D.III fighter. A nose radiator was then fitted along with modified flight controls, and on 1 December another test flight revealed greatly improved flight characteristics.

For the First Fighter Competition another new wing with aerodynamically-balanced ailerons was fitted. Manfred von Richthofen and other front-line pilots flew the J 7 and stated it had the best climb and speed among the inline-engined fighters. During the competition the J 7 survived two crash landings and was quickly repaired.

After the competition the J 7 was further modified and tested, and was then purchased as a trainer and demonstration aircraft. It was again demonstrated at the Second Fighter Competition along with the revised J 9, the D.I production prototype.

Despite its good performance and the quality of its engineering and construction, some prejudice remained against is low-wing configuration and all-metal construction.

Junkers J 7 Specifications

Engine:	160 hp Mercedes D.III	
Wing:	Span	9.20 m
	Area	11.7 m^2
General:	Length	6.70 m
	Height	2.60 m
	Empty Weight	656 kg
	Loaded Weight	805 kg
Maximum Speed:		205 km/h
Climb:	5000m	24.0 min

Left: The Junkers J 7 all-metal fighter prototype with original test radiator but new wings with conventional ailerons.

Below: The revised J 7 photographed on 29 March 1918 now has a nose radiator. It competed at the First Fighter Competition with wings similar to those shown. Note the robust pilot's headrest and turn-over structure.

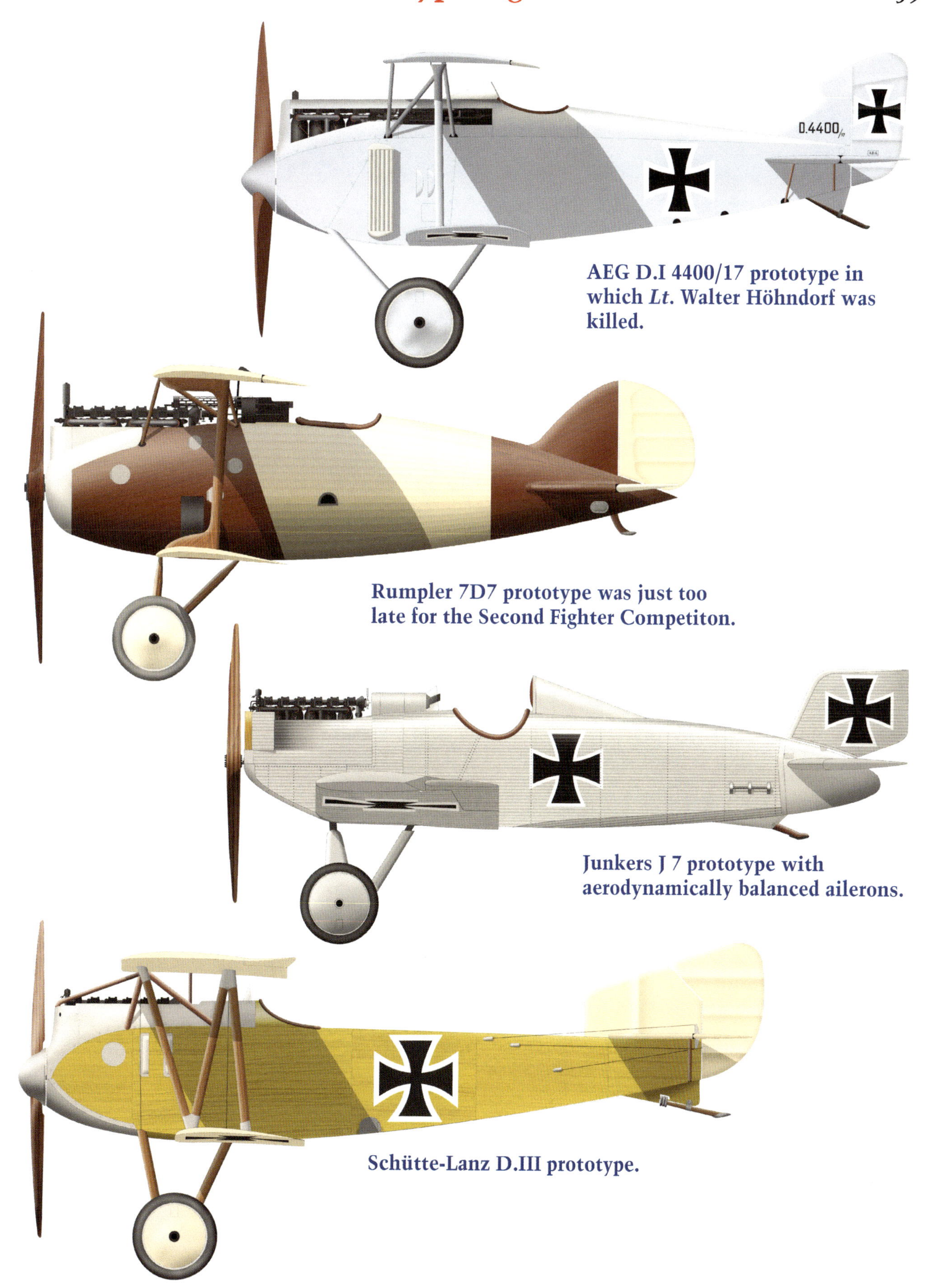

AEG D.I 4400/17 prototype in which *Lt.* Walter Höhndorf was killed.

Rumpler 7D7 prototype was just too late for the Second Fighter Competiton.

Junkers J 7 prototype with aerodynamically balanced ailerons.

Schütte-Lanz D.III prototype.

Schütte-Lanz at the First Fighter Competition

Schütte-Lanz was formed to build airships and later designed a number of airplanes, none of which were produced in quantity. The Schütte-Lanz D.III was designed to compete in the First Fighter Competition. Powered by a 160 hp Mercedes D.III, it was a conventional design that offered mediocre performance and was not selected for production or further development.

Schütte-Lanz D.III Specifications		
Engine:	160 hp Mercedes D.III	
Wing:	Span	8.00 m
General:	Length	6.50 m
	Loaded Weight	860 kg
Maximum Speed:		195 km/h
Climb:	1000m	3.0 min
	5000m	31.9 min

Above & Below: The Schütte-Lanz company primarily built airships, but then expanded into designing airplanes, none of which reached production. The Schütte-Lanz D.III fighter prototype was powered by the 160 hp Mercedes D.III engine. A neat, conventional design, when flown by von Arnim on 25 January it climbed to 1,000m in 3 minutes and 5,000m in 31.9 minutes, an unimpressive performance. Loaded weight was 1,980 pounds.

Rumpler at the First Fighter Competition

Development of the high-altitude Rumpler fighter began in early 1916 and continued throughout the war with a series of prototypes built.

Two versions of the Rumpler 7D4 fighter, the current prototype, competed at the First Fighter Competition. These demonstrated excellent climb and ceiling at the competition but were 'excessively touchy and very difficult to fly'. Not ready for production, their potential was obvious and development continued; *Idflieg* considered the design had enough potential to place a production order for 50 aircraft, D.1550–1599/18 in February 1918. Only 22 were completed before the armistice.

Rumpler D.I Specifications

Engine:	170 hp Mercedes D.IIIa	
Wing:	Span Upper	8.42 m
	Span Lower	7.38 m
	Chord Upper	1.32 m
	Chord Lower	1.09 m
	Gap	1.365 m
	Area	16.66 m²
General:	Length	5.75 m
	Height	2.56 m
	Empty Weight	615 kg
	Loaded Weight	805 kg
Maximum Speed (at 5000 m):		180 km/h
Climb:	5000m	23.7 min

Above & Below: Development of the high-altitute Rumpler fighter had begun in early 1916. Two Rumpler 7D4 fighter prototypes were tested at the First Fighter Competition; the parallel strut version above and the U-strut version below. Both were powered by the Mercedes D.IIIa engine. The photo above was taken 24 January 1918 at the start of the First Fighter Competition. (Lower photo from The Peter M. Bowers Collection/The Museum of Flight)

LVG at the First Fighter Competition

Well-known for its successful two-seat reconnaissance aircraft, LVG also designed a number of prototype fighters. The LVG E.I, first flown in 1915, was the earliest German two-seat fighter to have a fixed, synchronized gun for the pilot and a flexible gun for the observer. A monoplane, the first prototype was lost due to structural failure while being ferried to the front for operational evaluation. Investigation revealed that the screws holding the wing bracing struts had worked loose and the wings had collapsed. Despite its potential, development was abandoned and all subsequent LVG fighter designs were biplanes.

The LVG D10 prototype featured a very deep fuselage that filled the wing gap. The LVG D12, also known as the LVG D.II (the LVG D.I being the original designation for the Albatros D.II built under license) had a more normal configuration but the fuselage still filled the wing gap. The LVG D.III was tested in June 1917 but not produced.

The LVG D.IV was entered in the First Fighter Competition but was destroyed by an engine fire the first day of the competiton, eliminating it.

The D.V was a totally new and unsuccessful design with poor stability and controllability, probably due to its eccentric wing design, and was abandoned after a crash-landing in July 1918. The D.VI featured a more conventional wing cellule but entered flight testing just before the Armistice. No data are available for the D.V and D.VI, neither of which was entered in a fighter competition.

LVG D.IV Specifications

Engine:	185 hp Benz Bz.IIIbo V-8	
Wing:	Span	8.50 m
	Area	18.06 m^2
General:	Length	6.28 m
	Height	2.70 m
	Empty Weight	680 kg
	Loaded Weight	935 kg
Maximum Speed:		195 km/h
Climb:	5000m	28 min

Above: Two LVG D.IV prototypes were built. Both were lost due to engine failures, the first on 5 Jan. 1918 and the second on 29 Jan. 1918, the first day of the First Fighter Competition, when the engine caught fire and destroyed the airplane.

Above: The LVG D.V first flew in June 1918; its upper wing panels outboard of the center section pivoted to act as ailerons. Likely intended to compete in the Second Fighter Competition, it was fast but unstable with poor controllability. Development was abandoned after a crash-landing in July. Power was a 185 hp Benz Bz.IIIbm; no other data is available.

Below: The LVG D.VI, powered like the D.V with a 185 hp Benz Bz.IIIbm, first flew shortly before the Armistice. Clearly related to the D.V, its wing cellule was much more conventional. Again, no further data is available. Neither the D.V or D.VI had pleasing lines nor looked promising, and neither was entered in a fighter competition.

DFW at the Fighter Competitions

The most widely-produced German warplane of WWI was the DFW C.V, but DFW had no luck designing fighters. The first DFW fighter design was the eccentric T 28 *Floh* (Flea), with the gap between the wings completely filled by the fuselage. The *Floh* was fast for the time and power but *Idflieg* had no interest in it and development was abandoned.

DFW next designed the T 34 fighter in both biplane and triplane forms and entered them in the First Fighter Competition. However, both types were rejected for poor visibility from the cockpit before flight evaluations started and thus neither appeared in the official data tabulations, so no data is available for either type. The T 34-II triplane was further condemned for "unsuitable design", although details are lacking. Power for both was the Mercedes D.III.

DFW perservered with fighter development and the DFW D.I, powered by a Mercedes D.IIIa engine, was entered in the Second Fighter Competition. Again the DFW entry was rejected "for any front-line use" before flight evaluations began and again the type did not appear in the subsequent tables of data. However, the D.I was rebuilt in July and was promising enough that it underwent the full set of static load tests. However, the tests revealed structural weaknesses in the fuselage and tail and D.I development was terminated.

DFW D.I Specifications		
Engine:	170 hp Mercedes D.IIIa	
Wing:	Span Upper	9.08 m
	Area	23.00 m^2
General:	Length	5.50 m
	Empty Weight	639 kg
	Loaded Weight	819 kg
Maximum Speed:		177 km/h
Climb:	4000m	10 min

Above: The bizarre DFW T 28 *Floh* was fast for its 100 hp Mercedes D.I due to its excellent streamlining, achieving 180 km/h on its maiden flight, but *Idflieg* had no interest in the type.

Above: The DFW T 34-II triplane fighter prototype at the First Fighter Competition with the DFW T 34-I biplane fighter prototype behind it. The T 34-II was the only *new* triplane fighter design ever entered in a 1918 fighter competition.

Above: The DFW T 34-I biplane at the First Fighter Competition; power was a Mercedes D.III for both the biplane and triplane versions of the T 34. Both were rejected for poor view from the cockpit before flight evaluations began.

Below: The DFW D.I, powered by a Mercedes D.IIIa, was entered in the Second Fighter Competition. Like the T 34 prototypes it was rejected before the flight evaluations began, but the reasons are not known.

The Second Fighter Competition

German Fighter Aircraft Competing at the Second Fighter Competition, 27 May–28 June 1918

Aircraft	Engine	Remarks
Albatros D.X 2206/18 (#4914)	Benz Bz.IIIbo	
Albatros D.XI 2209/18 (#5045)	Siemens-Halske Sh.III	
Aviatik D.III 3550/18 (#10012)	Benz Bz.IIIbo	
Aviatik D.III (#10005)	Benz Bz.IIIbo	
Aviatik D.IV (#10008)	Benz Bz.IIIbv	Enlarged volume engine
Daimler D.I (#60)	Mercedes D.IIIbv	
DFW D.I	Mercedes D.IIIa	Rejected before flight evaluations began
Fokker V21 (#2310)	Mercedes D.IIIaü	High-compression engine
Fokker V23 (#2443)	Mercedes D.IIIa	
Fokker V24 (#2612)	Benz Bz.IVü	High-compression engine
Fokker V25 (#2732)	Oberursel Ur.II	
Fokker V27 (#2734)	Benz Bz.IIIboü	
Fokker V28 (#2735)	Oberursel Ur.II	Won the competition, proto. for Fokker E.V
Fokker V28 (#2735)	Oberursel Ur.III	
Fokker V28 (#2735)	Goebel Goe.III	
Fokker D.VII [#2268]	Mercedes D.IIIa	Standard operational type
Fokker D.VII (Alb.) 527/18 (#5148)	Mercedes D.IIIa	Standard operational type
Junkers J.9	Mercedes D.IIIaü	High-compression engine
Kondor D.I (#200)	Oberursel Ur.II	
Kondor D.II (#201)	Oberursel Ur.II	
Pfalz D.VIII 150/18	Siemens-Halske Sh.III (Rh)	Standard operational type
Pfalz D.VIII 158/18	Oberursel Ur.III	Standard operational type
Pfalz D.XII 1375/18	Mercedes D.IIIaü	High-compression engine
Pfalz D.XII 1387/18	B.M.W.IIIa	
Pfalz D.XIIa	Benz Bz.IIIoü	High-compression engine
Pfalz D.XIV	Benz Bz.IVü	High-compression engine
Roland D.VIb	Benz Bz.IIIaü	High-compression engine
Roland D.VII 224/18 (#3780)	Benz Bz.IIIbo	
Roland D.IX 3001/18 (#3900)	Siemens-Halske Sh.III	
Rumpler D.I 1552/18	Mercedes D.IIIa	
Rumpler D.I 1553/18 (#4402)	Mercedes D.IIIa	
Schütte-Lanz D.VII/3	Mercedes D.IIIaü	High-compression engine
Siemens-Schukert D.III 1627/18	Siemens-Halske Sh.III	
Siemens-Schukert D.III 1629/18	Siemens-Halske Sh.III	
Siemens-Schukert D.III 3008/18	Siemens-Halske Sh.III	
Siemens-Schukert D.IIIa 1622/18	Siemens-Halske Sh.III	
Siemens-Schukert D.V 7557/17	Siemens-Halske Sh.III	

Notes:

1. Numbers in parentheses are factory numbers.
2. Fokker D.VII #2268; the #2268 may be a factory or military number.
3. One of the Rumpler D.I fighters was also flown with a Mercedes D.IIIaü (high-compression).
4. The Fokker V.28 monoplane won the competition and, slightly modified, was placed into production as the Fokker E.V. The E.V was later re-designated the D.VIII.
5. Both Daimler (Mercedes) and Benz V-8 engines were built in geared and ungeared versions and their designations differed accordingly. The geared engines had the suffix 'm' for 'mit' (with), for with gears, while the ungeared engines had suffix 'o' for 'ohne' (without), for without gears, in their designations.

Fighter Evaluation Flights by Front-Line Pilots, Adlershof, July 1918

These flights were an adjunct to the Second Fighter Competition, and included some additional aircraft (noted by an ‡) that had not participated in the competition.

Aircraft	Engine	Remarks
Albatros D.X 2206/18 (#4914)	Benz Bz.IIIbo	
Albatros D.XII 2210/18 ‡	Mercedes D.IIIa	
Aviatik D.III 3550/18 (#10012)	Benz Bz.IIIbo	
Aviatik D.III (Versuch)(#10005)	Benz Bz.IIIbo	
Aviatik D.IV (#10008)	Benz Bz.IIIbv	
Aviatik D.VI ‡	Benz Bz.IIIbm	Evolved into D.VII & placed in production
Daimler D.I (#60)	Mercedes D.IIIbv	
Fokker V24 (#2612)	Benz Bz.IVü	
Fokker V27 (#2734)	Benz Bz.IIIboü	
Fokker V27 (#2734) ‡	Benz Bz.IIIbm	
Fokker V28 (#2735)	Oberursel Ur.II	
Fokker V28 (#2735)	Oberursel Ur.III	
Fokker D.VII ‡	B.M.W.IIIa	Standard operational type
Fokker D.VII ‡	Mercedes D.IIIa	Standard operational type
Fokker D.VII (Alb.) 527/18 (#5148)	Mercedes D.IIIa	Standard operational type
Fokker D.VII (Oaw.) ‡	Mercedes D.IIIa	Standard operational type
Junkers D.I	Mercedes D.IIIa	Evaluation batch ordered
Kondor D.I (#200)	Oberursel Ur.II	
Kondor D.II (#201)	Oberursel Ur.II	
Naglo D.II 1165/18 ‡	Mercedes D.IIIa	Only quadraplane entered in competitions
Pfalz D.VIII 150/18	Siemens-Halske Sh.III (Rh)	Standard operational type
Pfalz D.VIII 158/18	Oberursel Ur.III	Standard operational type
Pfalz Parasol (D.X) ‡	Siemens-Halske Sh.III	
Pfalz D.XII 1371/18 ‡	Mercedes D.IIIaü	High-compression engine
Pfalz D.XII 1387/18	B.M.W.IIIa	
Pfalz D.XIIa	Benz Bz.IIIoü	High-compression engine
Pfalz D.XIV ‡	Benz Bz.IVü	50 ordered, most cancelled
Roland D.VIb [2225]	Benz Bz.IIIav	
Roland D.VII 224/18 (#3780)	Benz Bz.IIIbo	
Roland D.VII ‡	Benz Bz.IIIbm	
Roland D.IX 3001/18 (#3900)	Siemens-Halske Sh.III	
Roland D.XIV 3002/18 ‡	Goebel Goe.III	
Rumpler D.I 1552/18	Mercedes D.IIIa	Evaluation batch ordered
Rumpler D.I 1553/18 (#4402)	Mercedes D.IIIa	
Schütte-Lanz D.VII/3	Mercedes D.IIIaü	High-compression engine
Siemens-Schukert D.III 1627/18	Siemens-Halske Sh.III	
Siemens-Schukert D.III 1629/18	Siemens-Halske Sh.III	
Siemens-Schukert D.IIIa 1622/18	Siemens-Halske Sh.III	
Siemens-Schukert D.IV 7555/17 ‡	Siemens-Halske Sh.III	
Siemens-Schukert D.V 7557/17	Siemens-Halske Sh.III	
Zeppelin D.I ‡	Mercedes D.IIIa	Evaluation batch ordered

Notes:

1. *Lt.* Reinhard (JGI) was killed 3 July 1918 when the top wing of the Zeppelin D.I he was flying tore off.
2. The Naglo D.II was the only quadraplane design submitted for evaluation during a fighter competition. Two modified Fokker Triplanes, a type then in production, flew at the First Fighter Competition, but only one *new* triplane design, the DFW T 34-II, was submitted for evaluation at a fighter competition. *Idflieg's* triplane craze was finally over, and now monoplanes were starting to appear, entered by Fokker, Junkers, and Pfalz. Of these designs the Fokker and Junkers designs were cantilever monoplanes and went into production. The Pfalz D.X monoplane with braced wing was abandoned.

Fokker at the Second Fighter Competition

Fokker arrived at the Second Fighter Competition with the most new fighter prototypes and the most momentum from the great combat success of the Fokker D.VII that was produced as a result of winning the First Fighter Competition.

In addition to a couple of production Fokker D.VIIs with Mercedes and BMW engines, Fokker brought variations on the Fokker D.VII theme as well as completely new airplanes.

The most significant of these new prototypes was the Fokker V28, a parasol-wing monoplane derived from the V17. The impressive performance of the V17 monoplane at the First Fighter Competition was impetus to design other monoplanes with better downward visiblity, and the V27 and V28 were those airplanes. In addition, the V25 was a low-wing monoplane derivative of the V17 designed to improve the downward view for the pilot.

Fokker V21

The Fokker V21, Works Number 2310, was the series prototype of the production Fokker D.VII. For the Second Fighter Competition it was powered by a high-compresion 180 hp Mercedes D.IIIaü engine. The exhaust exited the side of the cowling and the standard two machine guns were fitted in front of the cockpit. The V21 appeared in February 1918, and is seen below wearing the camouflage and markings of that time. Production D.VIIs had straight-sided crosses on their fuselage and rudder, although early production D.VIIs still had iron crosses on their upper wings. The specifications are for the original 160 hp Mercedes D.III engine.

Fokker V21 Specifications

Engine:	160 hp Mercedes D.III	
Wing:	Span	8.90 m
	Area	20.2 m^2
General:	Length	6.945 m
	Height	2.945 m
	Empty Weight	652 kg
	Loaded Weight	870 kg
Climb:	2000m	7.0 min
	3000m	12.0 min
	4000m	19.5 min
	5000m	27.0 min

Above: The Fokker V21 was a variant of the Fokker D.VII. The V22 was a D.VII version with 200 hp Austro-Daimler engine.

Fokker V23

The Fokker V23, Works Number 2443, was another mid-wing monoplane design similar to the V20 that had appeared at the First Fighter Competition. Again the design was rejected for the poor downward view afforded the pilot.

Fokker V23 Specifications		
Engine:	160 hp Mercedes D.III	
Wing:	Span	9.01 m
	Area	13.6 m^2
General:	Length	7.81 m
	Empty Weight	658 kg
	Loaded Weight	877 kg
Maximum Speed:		210 kmh
Climb:	1000m	2.5 min
	2000m	8.5 min
	3000m	12.0 min
	4000m	17.0 min
	5000m	23.0 min

Left, Below, & Bottom: The Fokker V23 was rejected because of the poor downward view for the pilot due to the wing location. Power was a 160 hp Mercedes D.III.

Fokker V24

The Fokker V24, Works Number 2612, was a version of the D.VII with a heavier, more powerful, over-compressed 200 hp Benz Bz.IVü engine. This same engine was also tested in the Pfalz D.XIV, a slightly enlarged derivative of the Pfalz D.XII. However, unlike Pfalz, the Fokker V24 had the same wings as the standard Fokker D.VII.

The additional weight of the Benz Bz.IVü offset the advantage of its greater power, and both performance and maneuverability with the Benz Bz.IVü were not as good as with the 185 hp BMW. IIIa. Combined with the need for the Bz.IV for two-seaters, this eliminated further consideration of the Benz engine for fighter use.

Fokker V24 Specifications		
Engine:	200 hp Benz Bz.IV	
Wing:	Span	8.90 m
	Area	20.2 m^2
General:	Length	7.14 m
	Height	3.15 m
	Empty Weight	750 kg
	Loaded Weight	989 kg
Climb:	2000m	6.0 min
	3000m	8.5 min
	4000m	12.5 min
	5000m	16.3 min

Above: The Fokker V24 was a variant of the Fokker D.VII powered by the over-compressed 200 hp Benz Bz.IVü engine. Although offering more power, the engine was also significantly heavier, and the 185 hp BMW.IIIa engine provided both better maneuverability and better performance due to its lighter weight.

Fokker V25

The Fokker V25, Works Number 2732, was derived from the V17 that had demonstrated such excellent performance at the First Fighter Competition. The V17 wing was mounted in the middle of the fuselage, and to improve the pilot's downward view the V25's wing was mounted at the bottom of the fuselage.

The V25 was tested with both the 110 hp Oberursel Ur.II and the more powerful 170 hp Goebel Goe.III engine. Even the 110 hp engine gave it good performance, but again the pilot's field of view was not considered satisfactory and it was not developed further.

Fokker V25 Specifications

Engine:	110 hp Oberursel Ur.II	
Wing:	Span	8.13 m
	Area	9.7 m^2
General:	Length	5.89 m
	Height	3.02 m
	Empty Weight	365 kg
	Loaded Weight	565 kg
Maximum Speed:		200 km/h
Climb:	1000m	1.5 min
	2000m	3.3 min
	3000m	6.5 min
	4000m	9.5 min
	5000m	13.3 min

Above & Below: The small Fokker V25 with 110 hp Oberursel Ur.II engine.

Above, Right, & Below: The Fokker V25 with 110 hp Oberursel Ur.II engine competed at the Second Fighter Competition but, like the earlier V17, despite good performance and maneuverability it was rejected due to the restricted downward field of view for the pilot. A robust crash pylon was built into the pilot's headrest to protect him in case of flipping over on landing.

Fokker V27

The Fokker V27, Works Number 2734, was another attempt to achieve the performance benefits of the monoplane configuration while solving the pilot's field of view problem that had led to rejection of the earlier Fokker monoplane fighter prototypes. In the V27, the wing was not mounted on the fuselage, which minimized aerodynamic drag, but was moved above the fuselage just above the pilot's eye level, giving the pilot exceptional fields of view both above and below the wing.

The V27 was powered by the experimental 200 hp Benz Bz.IIIbo ungeared V-8 engine. It had good performance but the experimental V-8 engine was not yet reliable enough for production and the V27 remained a prototype.

Interestingly, the Fokker V29, one of the winners of the Third Fighter Competition, used the same configuration as the V27; the main difference is that the V29 used the reliable, production BMW.IIIa engine instead on an experimental engine.

Fokker V27 Specifications		
Engine:	200 hp Benz Bz.IIIbo	
Wing:	Span	9.68 m
	Area	14.3 m²
General:	Length	6.34 m
	Height	3.0 m
	Empty Weight	602 kg
	Loaded Weight	840 kg
Maximum Speed:		200 km/h

Low drag was the key benefit of cantilever monoplane wings, and the parasol configuration combined that with exceptional field of view. The drawback was that the cantilever wing, even if strong enough to safely support the aerodynamic loads, was not as stiff in torsion as the biplane wing cellule with N-struts, with the result that aileron effectiveness was not as good as that of the biplane.

Above: The Fokker V27 parasol monoplane configuration solved the problem of the pilot's limited field of view suffered by the previous Fokker monoplane prototypes at the expense of greater drag from the struts supporting its wing. However, despite good performance its 200 hp Benz Bz.IIIbo experimental V-8 was not ready for production.

Fokker V28

Fokker V28 Specifications		
Engine:	110 hp Oberursel Ur.II	
Wing:	Span	8.34 m
	Area	10.7 m^2
General:	Length	5.865 m
	Height	2.82 m
	Empty Weight	360 kg
	Loaded Weight	560 kg
Maximum Speed:		200 km/h
Climb:	3000m	7.5 min
	4000m	10.5 min
	5000m	14.7 min
	6000m	19.5 min

With the Fokker V28, Works Number 2735, Fokker finally produced another design destined for production as the Fokker E.V. The V28 shared its parasol monoplane configuration with the V27, but was a smaller, lighter machine that was powered by a variety of rotary engines. Its parasol wing solved the pilot's field of view problem that had led to rejection of the earlier Fokker monoplane fighter prototypes. Like the V27, the wing was moved above the fuselage just above the pilot's eye level, giving the pilot exceptional fields of view both above and below the wing.

The V28 was the winner of the Second Fighter Competition despite its low-power 110 hp Oberursel Ur.II. It was fast and maneuverable, and had an exceptional field of view for the pilot.

The V28 was flown with at least four different rotary engines, including the 160 hp Oberursel Ur.III, the 160 hp Goebel Goe.III, and a Rhemag-built Sh.III of 160 hp. Of these only the Rhemag-built Sh.III would see quantity production, and that in limited numbers. So the Fokker E.V, the production version of the V28, was placed in production with the same low-power 110 hp Oberursel Ur.II used previously in the Fokker Triplane and Fokker D.VI. The E.V's performance led to curtailment of D.VI biplane production in favor of the faster E.V.

Above: The Fokker V28 parasol monoplane configuration solved the problem of the pilot's limited field of view suffered by the previous Fokker monoplane prototypes at the expense of greater drag from the struts supporting the wing. Smaller and lighter than its companion V-8 powered V27, the V28 was the prototype of the production Fokker E.V that went into production with the 110 hp Oberursel Ur.II previously used in the Fokker Triplane and Fokker D.VI biplane. Here the V28 is seen with the 11-cylinder Ur.III of 160 hp.

Fokker E.V/D.VIII

The Fokker V28 was the winner of the Second Fighter Competition and was placed into production as the Fokker E.V, unfortunately with the same low-power 110 hp Oberursel Ur.II used previously in the Fokker Triplane and Fokker D.VI.

The E.V was the minimum practical fighter aircraft that could be built with the available technology. It was strong, fast, and maneuverable, although its cantilever wings vibrated during high-G maneuvers and its aileron response was not as good as the Fokker biplanes with their torsionally stiffer wing cellules.

Production Fokker E.V fighters started arriving at the *Jastas* by 5 August 1918. Unfortunately, fatal in-flight wing failures followed on 16 and 19 August, and the E.V was immediately grounded.

The problem was traced to rot of the inner wing structure due to moisture condensation in the wing. The moisture had entered the wing via breathing holes and condensed. The problem was compounded by poor production quality control at the factory, which resulted in under-sized spars and faulty production procedures.

As a result, a new wing with strengthened spars was tested on 7 September and passed the structural tests. On 24 September Fokker was given permission to resume building the fighter, now re-designated the Fokker D.VIII. Fokker was also required to build new wings at his own expense for the 139 E.V fighters already completed. By the Armistice 80 D.VIII fighters had been accepted, but it is not known how many of these reached combat.

The results of the Third Fighter Competition showed that better fighters were available, and the D.VIII was likely destined for a relatively short career even had the war lasted into 1919.

Fokker E.V / D.VIII Specifications

Engine:	110 hp Oberursel Ur.II	
Wing:	Span	8.34 m
	Area	10.7 m^2
General:	Length	5.865 m
	Height	2.82 m
	Empty Weight	405 kg
	Loaded Weight	605 kg
Maximum Speed:		200 km/h
Climb:	1000m	2.0 min
	2000m	4.5 min
	3000m	7.5 min
	4000m	10.8 min
	5000m	15.0 min
	6000m	19.5 min

Above: The production Fokker E.V, powered by the 110 hp Oberursel Ur.II previously used in the Fokker Triplane and Fokker D.VI biplane, in service with *Jasta* 6, August 1918. The wingtips were more rounded than on the V28.

Fokker E.V

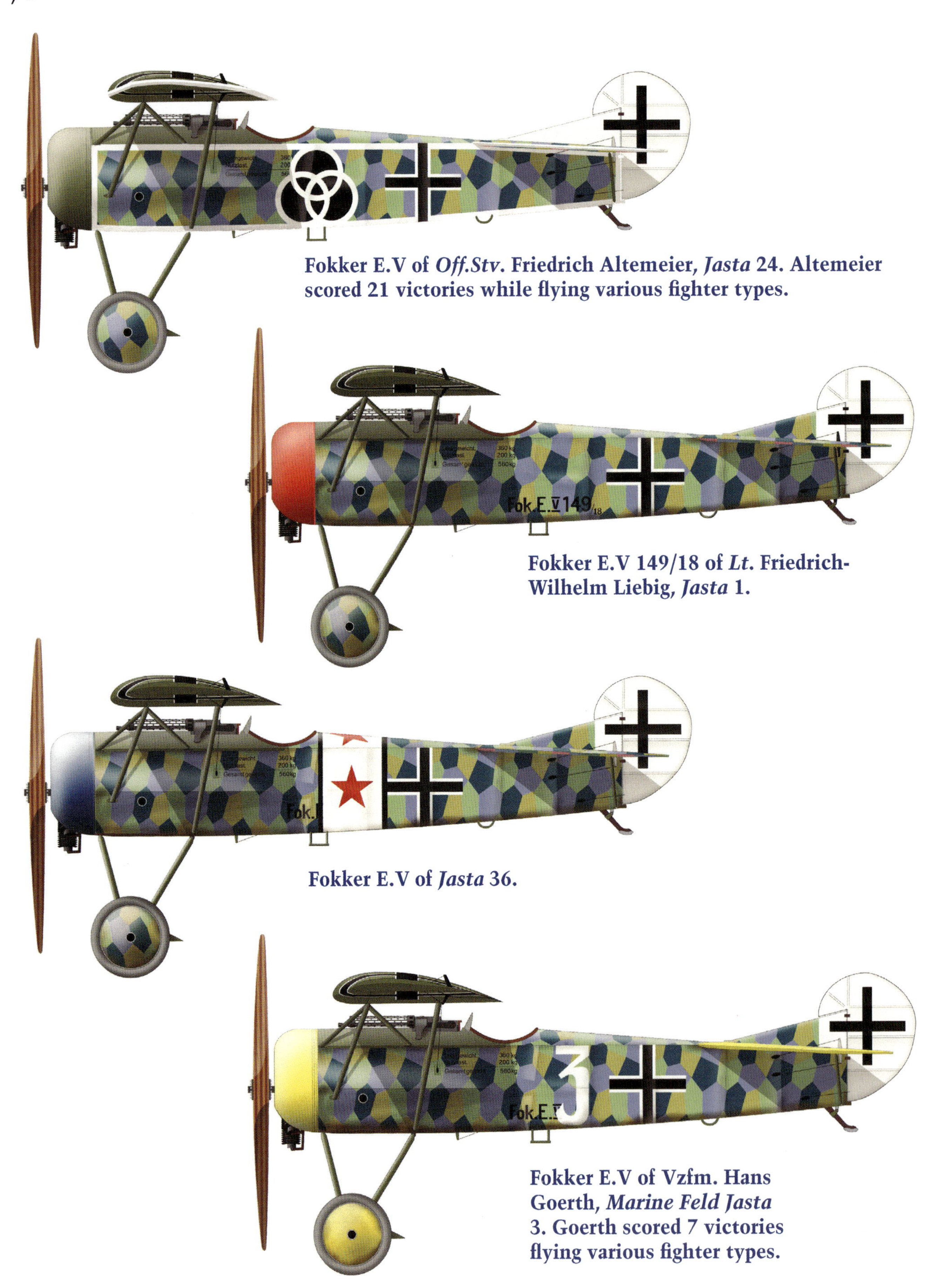

Fokker E.V of *Off.Stv*. Friedrich Altemeier, *Jasta* 24. Altemeier scored 21 victories while flying various fighter types.

Fokker E.V 149/18 of *Lt*. Friedrich-Wilhelm Liebig, *Jasta* 1.

Fokker E.V of *Jasta* 36.

Fokker E.V of Vzfm. Hans Goerth, *Marine Feld Jasta* 3. Goerth scored 7 victories flying various fighter types.

Pfalz at the Second Fighter Competition

The Pfalz company had gained some momentum in the fighter business from the showing of its D.VII at the First Fighter Competition. Moreover, by the Second Fighter Competition the Pfalz D.IIIa had been replaced in production by the improved D.XII. This was possible because the Pfalz company was sponsored by the German state of Bavaria, which wanted to retain as much autonomy as possible within the German Empire dominated by Prussia.

The stronger Pfalz D.VIII had been placed in production in lieu of the single-bay D.VII, and several D.VIII prototypes competed at the competion, as did the Pfalz D.XII. The Pfalz D.X Parasol monoplane was also present, but its wing design was not as innovative as Fokker's and it did not surpass the Pfalz D.VIII biplane using the same engine.

The Pfalz D.XIV, powered by the 200 hp Benz Bz.IVü, used a D.XII fuselage and a D.XII wing with span extended by one meter to support the greater weight of its engine. The extra power made it slightly faster than the Pfalz D.XII, but climb and maneuverability were not improved due to its greater weight. A small production order of 50 Pfalz D.XIV fighters was later cancelled after only a few had been built.

Above: Pfalz test pilot Gustav Bauer proudly stands before the prototype Pfalz D.X parasol monoplane in late 1918. Like the Pfalz Dr.I triplane and D.VII/D.VIII biplanes, it was powered by the 160 hp Siemens-Halske Sh.III counter-rotary engine. Although it did not officially compete in the Second Fighter Competition, it was one of the airplanes evaluated in July 1918 by front-line fighter pilots. Unlike the thick, cantilever Fokker wings, the Pfalz wing required extensive bracing, with its associated additional weight and drag. The Pfalz D.X was not developed further.

Pfalz D.XII

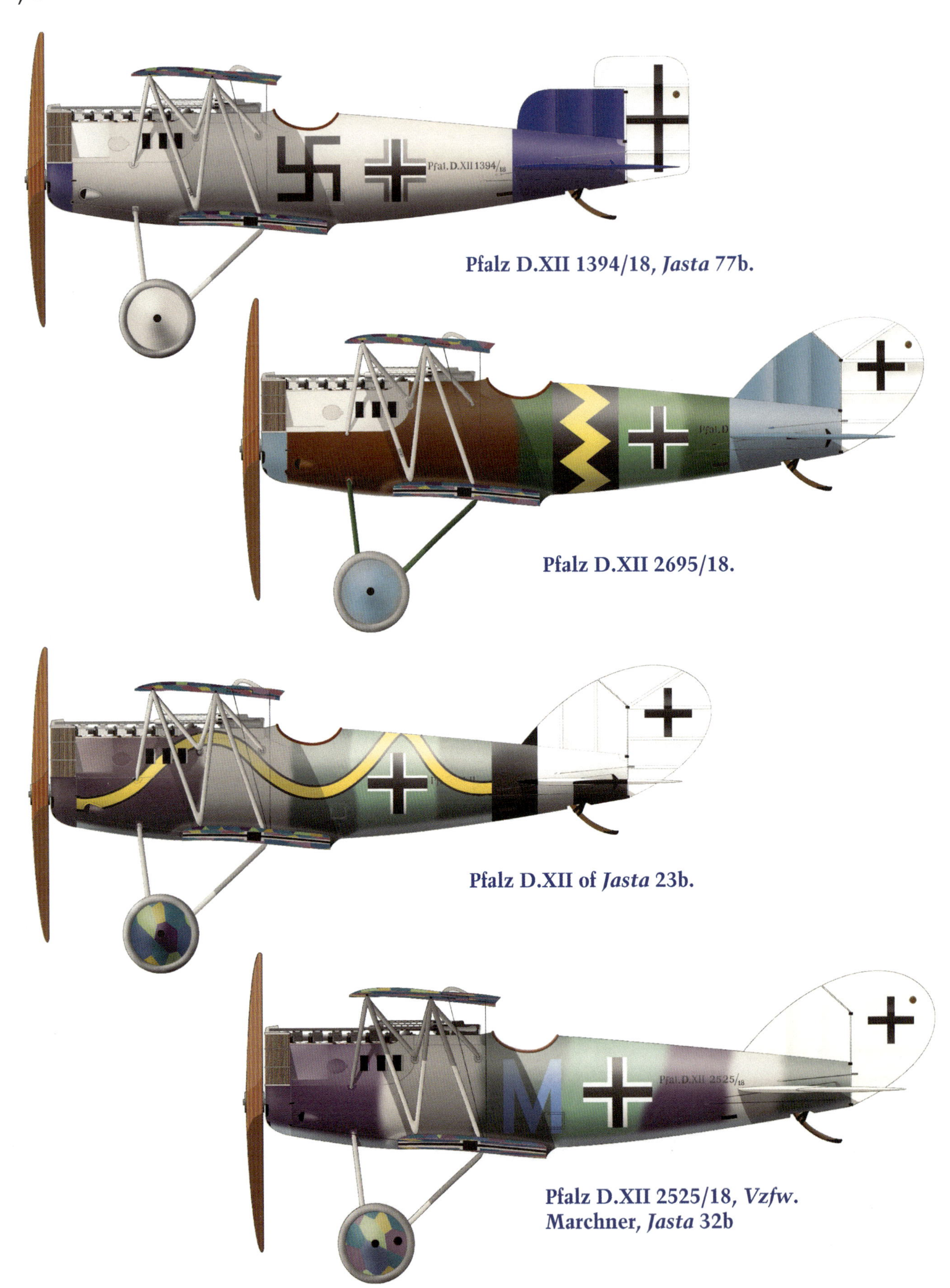

Pfalz D.XII 1394/18, *Jasta* 77b.

Pfalz D.XII 2695/18.

Pfalz D.XII of *Jasta* 23b.

Pfalz D.XII 2525/18, *Vzfw.* Marchner, *Jasta* 32b

Above: The German state of Bavaria wanted to maintain as much autonomy as possible within Imperial Germany and one manifestation of that desire was its support of the Pfalz company, located in Bavaria. Pfalz built airplanes throughout the war, initially license-built copies of the Morane-Saulnier L, and later original fighter designs. The Pfalz D.III was a good, solid fighter that just needed more power to be competitive with Allied fighters. The Pfalz D.XII looked similar to the Fokker D.VII and went into service in the summer of 1918; problems with its frontal, car-type radiator delayed its arrival by a couple of months. By the time the D.XII arrived the Fokker D.VII had a huge reputation and was in great demand by the pilots, most of whom wanted nothing else. The D.XII was another good, solid Pfalz design over-shadowed by its competition that never made a name for itself. But one manifestation of Bavaria's desire for autonomy exists today; the BMW company, which was created from the old Bavarian Rapp engine company in 1917. The Pfalz D.XII above is an early production example; later production aircraft had a rounded fin and rudder. Power was the Mercdes D.IIIa/av/avü series.

Left: The final production configuration of the Pfalz D.XII included a rounded fin and rudder.

Above: A prototype Pfalz D.VIII with N-struts and horn-balanced ailerons not used by the production version.

Pfalz D.XII Specifications

Engine:	170 hp Mercedes D.IIIa	
Wing:	Span Upper	9.00 m
	Span Lower	7.90 m
	Chord Upper	1.40 m
	Chord Lower	1.40 m
	Dihedral Lower	1 deg
	Gap	1.46 m
	Stagger	0.42 m
	Area	21.7 m^2
General:	Length	6.35 m
	Height	2.70 m
	Empty Weight	712 kg
	Loaded Weight	892 kg
Maximum Speed:		180 kmh
Climb:	1000m	3.5 min
	2000m	8 min
	3000m	14.5 min
	4000m	25.1 min

This Page: Powered by the 200 hp Benz Bz.IVü, the Pfalz D.XIV was an enlarged D.XII built using standard D.XII components with an extra meter of wingspan. It was about 5 km/h faster than the D.XII but climb and maneuverability were not improved. Due to the modest performance improvement and need for the Benz engine for two-seaters, the D.XIV received a production order for only 50 fighters that was later apparently cancelled.

Pfalz D.XIV Specifications

Engine:	200 hp Benz Bz.IVü	
Wing:	Span Upper	10.00 m
	Span Lower	9.15 m
	Chord Upper	1.40 m
	Chord Lower	1.40 m
	Dihedral Lower	1 deg
	Gap	1.47 m
	Stagger	0.38 m
	Area	25.43 m^2
General:	Length	6.32 m
	Height	2.70 m
	Loaded weight	1,032 kg
Maximum Speed:		190 kmh
Climb:	5000m	24.2 min

Albatros at the Second Fighter Competition

Above: Albatros D.X 2206/18, Work Number 4914, at the Second Fighter Competition. Power was from a Benz Bz.IIIbm geared V-8.

Below: The Albatros D.X used I-type interplane struts for low drag, but these tended to interfer with the pilot's field of view. The climb rate was too low for the D.X to be considered for production.

Albatros D.X Specifications

Engine:	195 hp Benz Bz.IIIbm V-8	
Wing:	Span Upper	9.90 m
	Span Lower	9.15 m
	Chord Upper	1.30 m
	Chord Lower	1.30 m
	Area	22.9 m²
General:	Length	6.18 m
	Height	2.74 m
	Empty Weight	666 kg
	Loaded Weight	905 kg
Maximum Speed		170 kmh

The First Fighter Competition had restored Fokker to fame and fortune, but had been a serious disappointment to Albatros. For more than a year Albatros fighters had been dominant in German fighter units and had been more numerous than all competitors combined. Fokker's triumph with the D.VII had ended Albatros's reign, and in fact Albatros now had to build the Fokker D.VII under license.

If Albatros had been somewhat complascent at the First Fighter Competition, they brought three new types to the Second Fighter Competition, including their first rotary-powered fighter, the D.IX.

The results were disppointing. The D.X had a mediocre rate of climb, and Henkel, the Albatros company test pilot, crashed the D.XI on landing during the competition. The D.XI was critized for bring prone to flipping over on landing, a legacy of its tall undercarriage needed to clear the propeller. Furthermore, the SSW D.III and Pfalz D.VIII, which used the same Sh.III engine as the Albatros D.XI, were already in production and had priority for the engine.

The Albatros D.XII was too late for the official trials for the Second Fighter Competition but was delivered in time for evaluation by the front-line pilots. It was judged inferior to both the Fokker D.VII and Pfalz D.XII and was returned to Albatros for continued development.

Albatros D.XI Specifications

Engine:	160 hp Siemens-Halske Sh.III	
Wing:	Span Upper	8.00 m
	Span Lower	6.60 m
	Chord Upper	1.50 m
	Chord Lower	1.00 m
	Stagger	0.36 m
	Area	18.5 m^2
General:	Length	5.58 m
	Height	2.87 m
	Empty Weight	494 kg
	Loaded Weight	689 kg
Maximum Speed:		180 kmh
Climb:	1000m	1.8 min
	2000m	4.3 min
	3000m	6.3 min
	4000m	9.3 min
	5000m	12.8 min
	6000m	17.8 min

Left: Front view of Albatros D.XI 2209/18, one of the competitors in the Second Fighter Competition. Power was from a Siemens-Halske Sh.III counter-rotary engine.

Right: Albatros D.XII 2210/18 shown here powered by the Mercedes D.III was evaluated in July, where it was judged slightly inferior to the Pfalz D.XII and was not chosen for production.

Above & Below: Two views of Albatros D.XI 2209/18, one of the competitors in the Second Fighter Competition. Power was from a Siemens-Halske Sh.III counter-rotary engine. Despite being flown by Albatros factory test pilot Henkel, this aircraft flipped over on landing at the competition after one mediocre altitude climb on June 10, 1918. The tall undercarriage legs needed for clearance for the two-bladed propeller made the airplane difficult to land safely. Slow aileron response resulted in the second D.XI prototype, 2208/18, being fitted with horn-balanced ailerons.

Prototype Albatros Fighters

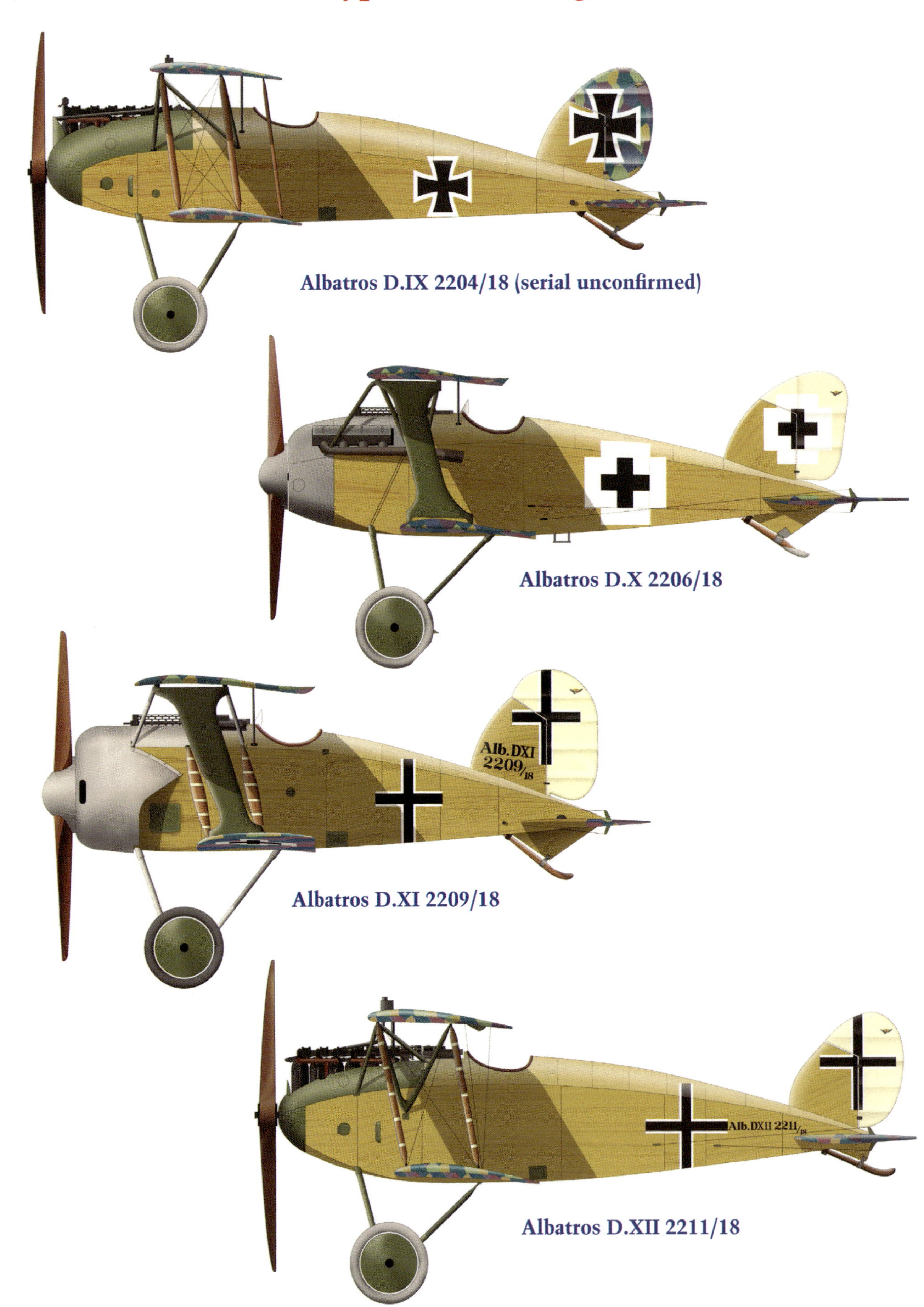

Albatros D.IX 2204/18 (serial unconfirmed)

Albatros D.X 2206/18

Albatros D.XI 2209/18

Albatros D.XII 2211/18

Schütte-Lanz at the Second Fighter Competition

Schütte-Lanz returned to the Second Fighter Competition with a new design, the D.VII powered by a 180 hp Mercedes D.IIIavü. Like the preceding D.III it was a conventional design of average performance and was not selected for production or further development.

Above, Right, & Below: The Schütte-Lanz D.VII, powered by a high-compression Mercedes D.IIIavü, competed at the Second Fighter Competition. A workman-like design with frontal radiator and N-struts, it had average performance. The thin wing profile required wire bracing and did not offer the benign stall characteristics of Fokker's thick, rounded wing.

Schütte-Lanz D.VII Specifications

Engine:	180 hp Mercedes D.IIIaü	
Wing:	Span	9.00 m
General:	Length	6.00 m
	Empty Weight	740 kg
	Loaded Weight	920 kg
Maximum Speed:		180 km/h
Climb:	1000m	2.4 min
	5000m	31.6 min

Aviatik at the Second Fighter Competition

Although not a stand-out performer at the First Fighter Competition, development of the Aviatik D.III fighter continued and it competed again at the Second Fighter Competition along with the Aviatik D.IV development of the D.III. The D.III continued working with the 195 hp Benz Bz.IIIb V-8, both geared and un-geared versions being tried. The D.IV differed in having the Benz Bz.IIIbv, a larger displacement version of the Bz.IIIbm.

Aviatik D.IV Specifications		
Engine:	195 hp Benz Bz.IIIbv V-8	
Wing:	Span	9.0 m
	Area	21.0 m^2

Below & Bottom: The Aviatik D.III fighter evaluated both the geared and un-geared versions of the Benz Bz.IIIb V-8 and competed at the Second Fighter Competition.

Above & Below: The Aviatik D.IV fighter competed at the Second Fighter Competition. The D.IV differed from the D.III in having the Benz Bz.IIIbv, a larger displacement version of the Bz.IIIbm used in the D.III. Both D.III and D.IV suffered from having pre-production engines and helping develop those engines may have been their primary purpose.

The Aviatik D.VI fighter, a new, two-bay design that owed little to preceding Aviatik fighters, was too late to compete at the Second Fighter Competition due to engine problems. However, it was available for evaluation by the front-line pilots after the official phase of the competition. The D.VI was powered by the 195 hp Benz Bz.IIIbm geared V-8 and proved to have excellent flight characteristics, but the single example built was overtaken by the similar Aviatik D.VII.

Aviatik D.VI Specifications

Engine:	195 hp Benz Bz.IIIbm V-8	
Wing:	Span	9.66 m
General:	Length	6.10 m
	Height	2.50 m
	Empty Weight	750 kg
	Loaded Weight	940 kg
Maximum Speed:		188 kmh
Climb:	5000m	17.8 min

Above & Below: The Aviatik D.VI fighter was too late to compete at the Second Fighter Competition but was in time to be evaluated by the front-line pilots in July. Its flying qualities were excellent, but it was overtaken by the similar Aviatik D.VII.

Above & Below: Two views of the Aviatik D.VI, which was evaluated by pilots in conjunction with the Second Fighter Competition despite not being a formal entry in the competition. The example at bottom has a larger fin and rudder, likely as a result of flight testing. Power was from a Benz Bz.IIIbm geared V-8.

Naglo at the Second Fighter Competition

Built by the Naglo Werft of Pichelsdorf near Berlin, apparently the Naglo D.II quadraplane was designed on the basis of 'more is better'. The retrogressive quadraplane wing cellule was added to a fuselage derived from the Albatros D.V. The Naglo D.II was evaluated in conjunction with the Second Fighter Competition and debriefing notes indicate it was to be modified and re-evaluated. Powered by a Mercedes D.III engine, it was intended to carry the standard German fighter armament of two machine guns. Official type testing was done 24 May 1918, and evaluation pilots praised the contruction and workmanship but wanted improved flying qualities. Designed by *Ing*. Gnädig, who was employed by Albatros at the time, its appearance resulted in his termination for conflict of interest and the prototype was rejected at the request of Albatros. The pylon for attaching the lowest wing was built into the fuselage. No performance data has survived.

Naglo D.II Specifications

Engine:	160 hp Mercedes D.III	
Wing:	Span	9.00 m
	Area	22.40 m^2
General:	Empty Weight	914 kg

Above & Below: The Naglo D.II fighter prototype, serial 1161/18, was evaluated in conjunction with the Second Fighter Competition. The Naglo was the only *new* design at any of the fighter competitions with more than two wings.

SSW at the Second Fighter Competition

The SSW D.III had made a good showing at the First Fighter Competition and SSW returned to the Second Fighter Competition with a number of D.III variations plus a D.IV and a D.V, the D.V being a two-bay version of the D.IV. The different versions of the D.III represented on-going development of the type for the best performance and maneuverability.

The D.IV proved superior to both the earlier D.III, which had a larger upper wing, and the two-bay D.V. In fact, two of the three D.V prototypes were rebuilt to single-bay D.IV configuration. Several of the SSW fighters at the competition had high-compression versions of the Siemens-Halske engine, the Sh.IIIa. The competition confirmed the excellent performance and high-altitude manueverability of the SSW D.IV and *Idflieg* ordered more as a result.

SSW D.IV Specifications

Engine:	160 hp Siemens-Halske Sh.III	
Wing:	Span Upper	8.350 m
	Span Lower	8.350 m
	Chord Upper	1.000 m
	Chord Lower	1.000 m
	Area	15.12 m^2
General:	Length	5.580 m
	Height	2.700 m
	Empty Weight	540 kg
	Loaded Weight	738 kg
Maximum Speed:		184 km/h
Climb:	1000m	1.9 min
	2000m	3.7 min
	3000m	6.4 min
	4000m	9.1 min
	5000m	12.1 min
	6000m	15.5 min
Ceiling:		~ 8000m

SSW D.V Specifications

Engine:	160 hp Siemens-Halske Sh.III	
Wing:	Span	8.86 m
General:	Length	5.70 m
	Empty Weight	514 kg
	Loaded Weight	734 kg
Maximum Speed:		184 km/h
Climb:	1000m	1.8 min

Above: SSW D.III in service at *KEST* 4b.

Right: Although the SSW D.III offered exceptional climb and ceiling, pilots wanted more speed. The D.III was modified into the similar D.IV by reducing the upper wing chord to that of the lower wing for reduced weight and drag; speed was improved by 10 kmh at the cost of a very slight reduction in climb rate. Combat pilots rated the SSW D.IV as superior to all other fighters, Allied or German, but it was not nearly as easy to fly as the Fokker D.VII.

Right: Here SSW D.IVs are in *Jasta* service. The interplane struts taper closer together on the upper wing than on the D.III due to the reduced chord. The SSW D.IV had exceptional climb and maneuverability but was not as fast as the Fokker D.VII with BMW engine.

Kondor at the Second Fighter Competition

Despite making no impression at the First Fighter Competition with its D.I, which was present but did not compete, Kondor became more active in designing fighters and fielded two D.II prototypes, Work Number 200 and Work Number 201, for the Second Fighter Competition. Again, the designations were for convenience and not official *Idflieg* designations.

The Kondor D.II was derived from the earlier D.I by replacing the single-spar lower wing with a new, stronger two-spar wing fitted with ailerons. The revised wing cellule was stronger, gave better performance, and better roll rate for excellent maneuverability. Both Kondor D.II prototypes were powered by the 110 hp Oberursel Ur.II rotary. They were the smallest fighters at the Second Fighter Competition.

At the competition the Kondor D.II was assessed as having very fine flight characteristics but poor performance, especially the mediocre climb, and there was no chance of a production order.

Kondor D.II Specifications

Engine:	110 hp Oberursel Ur.II	
Wing:	Span	7.59 m
	Wing Area	13.34 m^2
General:	Length	4.87 m
	Height	2.41m
	Empty Weight	380 kg
	Loaded Weight	560 kg
Maximum Speed:		175 km/h
Climb:	3000m	10.4 min
	5000m	30.5 min

The Kondor D.VI, which did not appear at any of the fighter competitions, is included for continuity. It was an interesting attempt to improve the pilot's field of view forward and upward by completely eliminating the wing center section. The D.VI was clearly related to the earlier D.II, but the revised wing design was structurally and aerodynamically problematic. In particular, induced drag from wingtip vortices from the upper wing was twice that of a normal design, which limited its speed and climb.

Above: The Kondor D.II was powered by the 110 hp Oberursel U.II and had a conventional wing cellule. By this time 110 hp was generally not enough power to be competitive with the 200 hp SE5a and the 200–220 hp Spad 13. More powerful rotary engines were in development but none were truly ready for production.

Above: Like the D.I, the Kondor D.II was powered by a 110 hp Oberursel U.II. However, the Kondor D.II had stronger, two-spar lower wings.

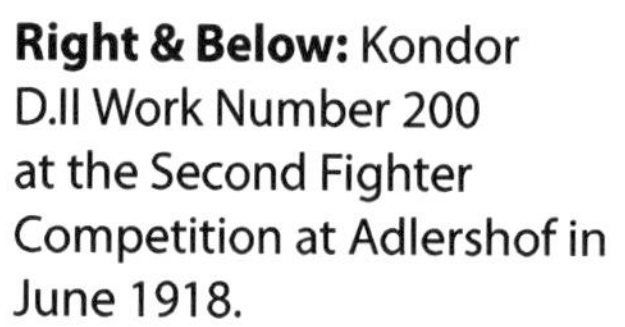

Right & Below: Kondor D.II Work Number 200 at the Second Fighter Competition at Adlershof in June 1918.

Above & Below: Like the Kondor D.II, the Kondor D.VI was powered by the 110 hp Oberursel U.II. The D.VI was an attempt to improve the pilot's field of view forward and upward by removing the center section of the upper wing. The D.VI undoubtedly created more induced drag (drag induced by lift from the wingtip vortices) than a conventional biplane because of the missing center section, which would have reduced its performance noticeably compared to its D.II predecessor. The D.VI was too late for the Second Fighter Competition and was abandoned before the Third Fighter Competition.

Kondor D.VI Specifications

Engine:	145 hp Oberursel Ur.III	
Wing:	Span	8.25 m
	Wing Area	13.80 m^2
General:	Length	5.80 m
	Height	2.41m
	Empty Weight	420 kg
	Loaded Weight	645 kg
Maximum Speed:		170 km/h

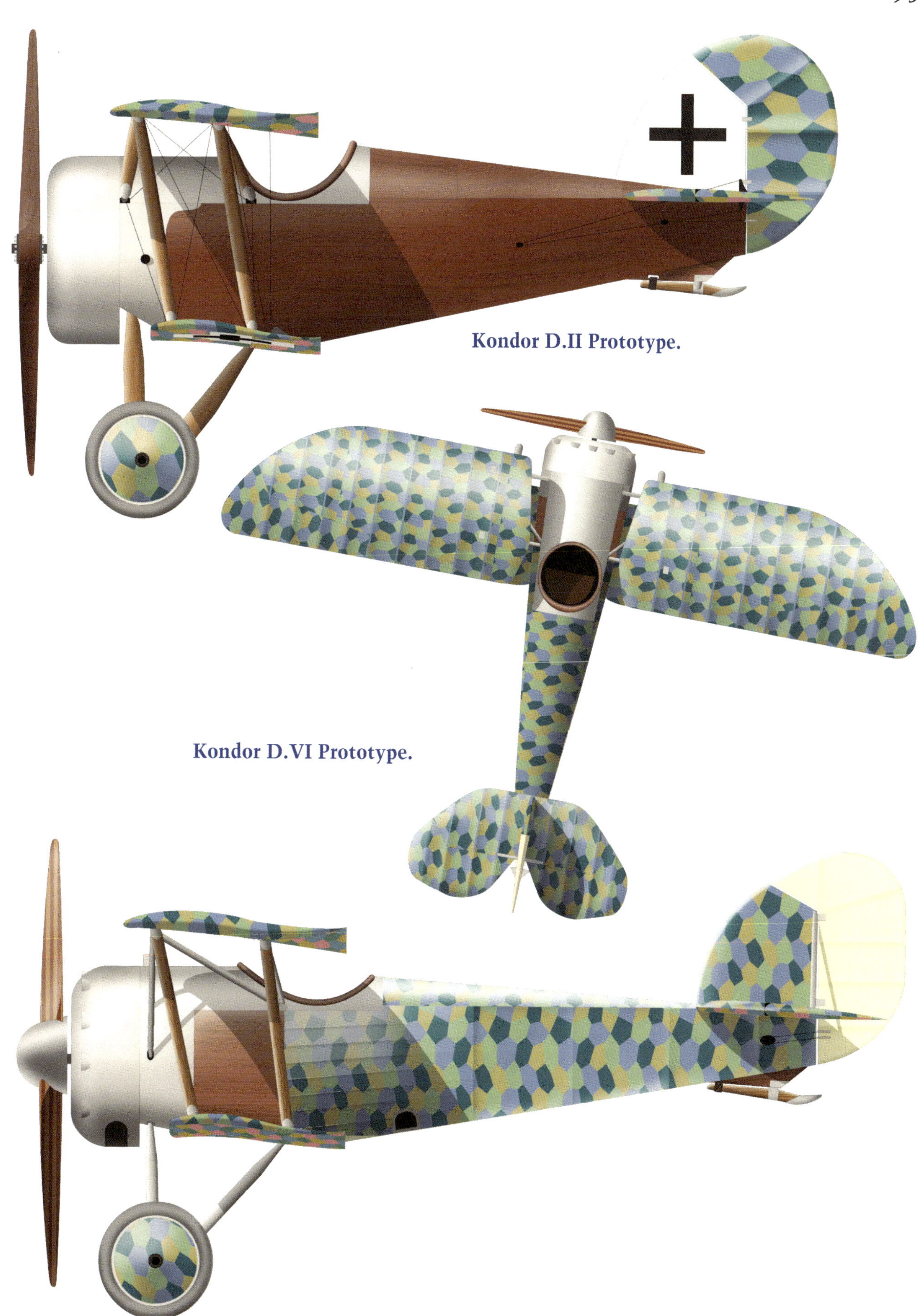
Kondor D.II Prototype.
Kondor D.VI Prototype.

Prototypes at the Second Fighter Competition

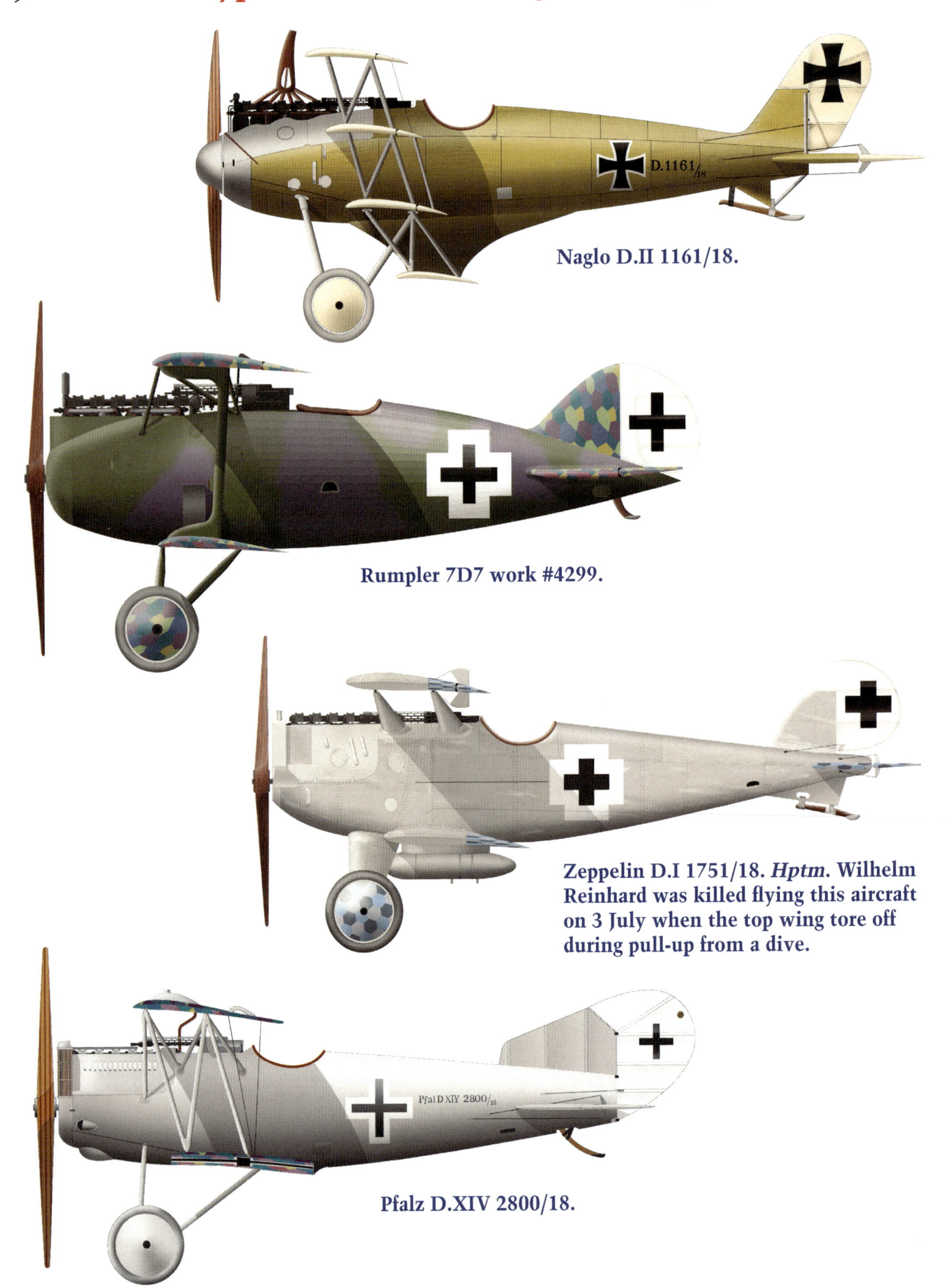

Naglo D.II 1161/18.

Rumpler 7D7 work #4299.

Zeppelin D.I 1751/18. *Hptm.* Wilhelm Reinhard was killed flying this aircraft on 3 July when the top wing tore off during pull-up from a dive.

Pfalz D.XIV 2800/18.

Daimler at the Second Fighter Competition

The Daimler-Motoren-Gesellschaft AG was the largest engine manufacturer in Germany and was well-known for its aircraft engines using the trade name Mercedes. In addition to aero-engine manufacturing, Daimler repaired airplanes, built more than 200 Friedrichshafen bombers under license, and also designed and built prototype combat aircraft, eventually including fighters.

In September 1917 the prototype 185 hp Mercedes D.IIIb V-8 aircraft engine, based on the Hispano-Suiza, was running well on the test stand, and the company decided to build a prototype fighter to flight-test the engine. The Daimler D.I began flight tests at the end of November with the geared version of the engine, the D.IIIbm. Initially it was tail heavy, but by January that was corrected. However, climb rate was still not satisfactory, and it was not until the Second Fighter Competition that the D.I was ready to compete. There it demonstrated an average climb time to 6,000 meters of 28.5 minutes, a time matched by the competing Aviatik D.IV powered by the 195 hp Benz Bz.IIIbm geared V-8. Front-line pilots did not like the Aviatik, and the Daimler V-8 engine was the Daimler D.I's main advantage. *Idflieg* ordered 20 Daimler D.I fighters on 23 July, primarily to speed development of the engine. One redesigned D.I was ordered on 22 September.

Daimler next built a D.II prototype with thick wings and wireless cellule, likely inspired by the Fokker D.VII. The D.II flew in July, too late for the Second Fighter Competition. In late August the D.II flew in comparison with a Fokker D.VII; the D.II proved equal in speed and had a much better climb rate. D.II development continued to the Armistice.

Above & Left: The Daimler D.I fighter prototype was powered by Daimler's Mercedes D.IIIbm geared V-8 engine and demonstrated a good climb rate at the Second Fighter Competition. German V-8 engines entered production too late to power any aircraft that reached the front.

Daimler D.I Specifications

Engine:	185 hp Mercedes D.IIIb V-8	
Wing:	Span	9.90 m
	Area	22.6 m^2
General:	Length	7.30 m
	Height	2.76 m
	Empty Weight	750 kg
	Loaded Weight	925 kg
Maximum Speed		183 kmh
Climb:	6000m	30 min

Daimler D.II Specifications

Engine:	185 hp Mercedes D.IIIb V-8	
Wing:	Span	9.00 m
	Area	22.32 m^2
General:	Length	7.20 m
	Height	2.60 m
	Empty Weight	742 kg
	Loaded Weight	990 kg
Maximum Speed		190 kmh

Above & Below: Another Daimler D.I fighter prototype was rolled out in March 1918; this version had shorter exhaust pipes and wore the new national insignia, but was still powered by Daimler's Mercedes D.IIIbm geared V-8 engine. Development of this prototype continued to improve engine cooling and aerodynamic issues, and it competed at the Second Fighter Competition, where its main asset was its engine.

Above, Right, & Below: The Daimler D.I fighter prototype as it appeared when it competed at the Second Fighter Competition. The most impressive D.I asset was its engine, and 20 aircraft were ordered on 23 July 1918 to help speed development of the promising engine to get it into production. The engine eventually went into production, but unfortunately too late to power any fighters in combat.

Above & Below: Yet another Daimler D.I fighter prototype was ordered on 22 September1918 as D.8800/18. This version was still powered by Daimler's Mercedes D.IIIbm geared V-8 engine but discarded the airfoil radiator of earlier D.I prototypes for a radiator in the nose. The rudder and fin appear to be taller than previous prototypes.

This Page: The Daimler D.II fighter prototype had thick, cantilever wings and initially flew without interplane struts, but is seen here after they were installed. The D.II arrived too late for the Second Fighter Competition, but in late August the D.II was flown with a Fokker D.VII for comparison. The D.II proved equal in speed and had a much better climb rate, but the type of engine in that Fokker is not known. D.II development continued to the Armistice. Its V-8 engine finally went into production, but unfortunately too late to power any fighters in combat.

Above & Below Right: The Daimler L11 fighter prototype was powered by Daimler's Mercedes D.IIIbm geared V-8 engine; here it is seen in its final configuration with vane-balanced ailerons.

Bottom: The Daimler L11 as delivered with unbalanced ailerons. It had a top speed of 240 km/h, very fast for its time.

Daimler's final single-seat fighter, the L11, was a parasol monoplane that apparently first flew on October 1, 1918. It would have been designated the Daimler D.III had it been accepted by *Idflieg*, but was too late for that. Like the other Daimler fighters, the L11 used the 185 hp Mercedes D.IIIb V-8 engine. The L11 proved to have exceptional performance, with a top speed of 240 km/h and a climb to 5,000 meters in 13 minutes. The L11 was delivered with unbalanced ailerons and was then modified with vane-balanced ailerons that received a patent in 1919.

Junkers at the Second Fighter Competition

The debut of the all-metal Junkers D 7 single-seat fighter at the First Fighter Competition had been successful and *Idflieg* awarded Junkers an open contract for 100 all-metal aircraft, the first 20 of which were to be D.I fighters.

Junkers began work on the J 9, the production prototype for the D.I fighter, in February 1918. Changes from the J 7 were generally minor and were done to make the J 9 easier to build in quantity. First flight of the J 9/I is thought to have been on 12 May.

For the Second Fighter Competition a high-compression Mercedes D.IIIaü was installed for better high-altitude speed and climb. During the climb competition only the Rumpler D.I had better climb than the J 9 among the inline-engine fighters.

The J 9/II, powered by an experimental 195 hp Benz Bz.IIIbo V-8, appeared in June, but engine problems prevented its planned appearance at the Second Fighter Competition and performance information is not available.

Despite the J 9's excellent flight performance and flying qualities, pilots still did not like the low-mounted wing due to its limited downward visibility. The combat pilots expressed a strong preference for a biplane or parasol monoplane configuration. However, *Idflieg* ordered ten more D.I production fighters for combat evaluation, realizing that only combat would determine the value of the low-wing configuration. Ironically, the compelling aerodynamic and structural benefits of the low-mounted cantilever wing made it the nearly universal configuration of choice for fighter designers in the world war to follow a generation later, showing that the experts are not always right! But the parasol configuration was still viable in 1918.

Junkers J 9/I Specifications

Engine:	180 hp Mercedes D.IIIaü	
Wing:	Span	9.20 m
	Area	14.8 m²
General:	Length	6.70 m
	Height	2.60 m
	Empty Weight	655 kg
	Loaded Weight	835 kg
Maximum Speed (180 hp Mercedes D.IIIa):		176 km/h
Climb:	1000m	3.0 min
	2000m	5.8 min
	3000m	9.5 min
	4000m	15.3 min
	5000m	24.6 min

Above: The Junkers J 9/I prototype for the D.I production fighter was entered in the Second Fighter Competition. It was flown by *Lt.* Krohn on June 6, 10, and 14 at Adlershof and became the world's first all-metal production fighter. It is in military camouflage and markings and the machine guns have been fitted. For better rearward visibility the roll-over structure is open and much smaller than that of the earlier J 7 prototype.

Zeppelin at the Second Fighter Competition

The Zeppelin company is normally thought of in conjunction with airships. In fact, large, rigid airships are often called 'Zeppelins' regardless of manufacturer, such was the prominence of the Zeppelin company's rigid airships. However, Zeppelin also built Giant bombers at its Staaken facility and designed and built giant, all-metal flying boats at its Lindau facility. These all-metal flying boats were designed by *Dipl.-Ing.* Claude Dornier, later to become well-known as founder of the Dornier aircraft company before WW2.

Dornier was also the designer of the Zeppelin D.I all-metal fighter, sometimes known as the Dornier D.I after its designer. The Junkers company built the world's first all-metal production aircraft, the Junkers J.I, and also the world's first all-metal production fighter, the Junkers D.I. But the Zeppelin D.I was the world's first *stressed-skin* all-metal fighter, or fighter in which the metal skin served as an integral, load-bearing part of the structure. The Zeppelin D.I was the most structurally-advanced airplane to fly in World War One, and its structure was similar to that of modern all-metal aircraft.

The proposal for the Zeppelin D.I was approved by *Idflieg* on 28 February 1918. Six prototypes were ordered, D.1750–1755/18, three to be powered by the 170 hp Mercedes D.IIIa and three by the experimental 195 hp Benz Bz.IIIb V-8 then in development. However, the V-8 was never installed in a Zeppelin D.I, the BMW.IIIa being used instead.

One June 4, 1918 the first prototype, D.1752/18, was taken for its first flight by test pilot *Vzfw*. Heinz Ruppert. The second D.I prototype, D.1751/18, competed a the Second Fighter Competition. It was flown by a number of pilots, then *Hptm*. Wilhelm Reinhard, Commander of *JG*I since the death of the Red Baron and a 20-victory ace nominated for the *Pour le Mérite*, took it up on 3 July. Reinhard put the prototype D.I through its paces, and while pulling out of a steep dive the top wing tore off and Reinhard

Zeppelin D.I Specifications

Engine:	170 hp Mercedes D.IIIa	
Wing:	Span	7.80 m
	Area	16.66 m^2
General:	Length	6.40 m
	Height	2.60 m
	Empty Weight	725 kg
	Loaded Weight	883 kg
Climb:	5000m	25 min

Above: The all-metal Zeppelin D.I 1752/18, the first protoype to fly, is seen here. Zeppelin D.I 1751/18, the second prototype to fly, was entered in the Second Fighter Competition where, during an evaluation flight on July 3, 1918, the upper wing tore off. *Lt.* Wilhelm Reinhard, who had succeeded Manfred von Richthofen as the commander of *JGI*, was killed in the subsequent crash. *Lt.* Hermann Göring had flown the fighter immediately prior to *Lt.* Reinhard.

was killed in the ensuing crash. This tragedy should not have happened; the D.I had not yet passed its official static load tests and regulations prohibited front-line pilots from flying aircraft until all load tests were passed.

Regardless of the tragedy, the combat pilots who had flown the Zeppelin D.I all agreed that it was on average superior to the other fighters powered by a Mercedes engine. The D.I was said to have flown tighter turns than the Fokker D.VII with about equal maneuverability. In addition to its maneuverability, the D.I had good speed, very good rate of climb, and excellent flying qualities.

The Zeppelin's obvious potential was such that development was continued. After the wings and their attachments had been reinforced, airframe D.1750/18 was tested to destruction from late September to 11 October and easily passed the load test requirement, which had been raised to 6.5 Gs from 5 Gs after Reinhard's fatal accident.

Above: The all-metal Zeppelin D.I 1752/18, the first protoype to fly, displays its clean, innovative lines.

Below: A Zeppelin D.I with 185 hp BMW.IIIa being tested by the U.S. Army at McCook Field in 1922.

Roland at the Second Fighter Competition

Roland made a reasonable showing at the First Fighter Competition with its D.VI, which was placed into small-scale production. Unfortunately for Roland, that was the high point of Roland success in the fighter competitions.

Roland brought a greater variety of prototypes to the Second Fighter Competition. In addition to Roland D.VIb production aircraft, Roland brought a revised D.VII and a redesigned D.IX, both revisions of aircraft that had competed in the First Fighter Competition. The new addition to the Roland stable was the D.XIV, a fighter resembling the D.IX but powered by the experimental 160 hp Goebel Geo.III rotary in place of the Siemens-Halske rotary of the same power.

Roland D.XIV Specifications	
Engine:	160 hp Goebel Goe.III

The results could only have been disappointing. The D.VIb remained in small-scale production but the engine in the D.VII was still not ready for production, resulting in termination of the D.VII. The revised D.IX had no special advantage over the similarly-powered Pfalz D.VIII and SSW D.III that were already in limited production and service, and the D.IX program was also terminated. The final blow was termination of the Roland D.XIV due to serious problems with its experimental engine.

Right: The Roland D.VIa, now in small-scale production, participated in the Second Fighter Competition. Despite the superiority of the Fokker D.VII, the Roland D.VI remained in limited production until the war's end, the only bright spot for Roland in 1918. Here a Roland D.VIa is flown postwar.

Left: The revised Roland D.IX 3001/18 fighter prototype powered by the 160 hp Siemens-Halske Sh.III was entered in the Second Fighter Competition. With no advantage over the similarly-powered SSW D.III and Pfalz D.VIII, both already in production, the Roland D.IX program was terminated.

Above & Below: The Roland D.XIV 3003/18 fighter prototype powered by the 160 hp Goebel Goe.III was entered in the Second Fighter Competition. Problems with the experimental engine dogged the aircraft, limiting its flights and eliminating it from competition. The D.XIV was very similar to the D.IX powered by the 160 hp Siemens-Halske Sh.III.

Rumpler at the Second Fighter Competition

Rumpler's situation at the Second Fighter Competition was mixed. Rumpler had been working on its high-altitude D.I fighter since early 1916, and prototypes had given good performance at the first fighter competition, leading to a production order for 50 fighters. That was the good news.

The other news was that the design was not firming up into a practical production fighter. The next prototype in the D.I sequence, the Rumpler 7D7, started its type-test process in February. A number of structural weaknesses had to be corrected, and the testing was prolonged as failures occurred and fixes were applied. A report issued on May 8, more than two months after testing started, still listed a number of items that needed strengthening before series production could start.

In trying to get extreme performance from the D.I, Rumpler engineers had calculated component strength to the limit to achieve minimum weight. Rumpler test pilot Gustav Basser recounted that the tail twisted during flight maneuvers, a quality that would hardly inspire confidence in combat pilots dependent on their fighter's robust airframe to survive high-G maneuvering combat. In fact, this weakness necessitated a new, more conventional – and heavier – fuselage design. Flight qualities were also proving unacceptable, and *Idflieg* required the design be rebuilt before release to front-line units could be considered. Development continued...

Rumpler D.I Specifications

Engine:	170 hp Mercedes D.IIIa	
Wing:	Span Upper	8.42 m
	Span Lower	7.38 m
	Chord Upper	1.32 m
	Chord Lower	1.09 m
	Gap	1.365 m
	Area	16.66 m^2
General:	Length	5.75 m
	Height	2.56 m
	Empty Weight	615 kg
	Loaded Weight	805 kg
Maximum Speed (at 5000 m):		180 km/h
Climb:	5000m	23.7 min

Above: Two Rumpler D.I fighter prototypes were entered in the Second Fighter Competition. Again, both were powered by the Mercedes D.IIIa engine. D.I 1589/18 above was photographed postwar so the guns have been removed.

The Third Fighter Competition

German Fighter Aircraft Competing at the Third Fighter Competition, October 10–28, 1918

This Competition was officially restricted to fighters with B.M.W. IIIa engines, but a number of rotary-engined fighters were included for comparison. The competition was held from 10 October to 2 November.

Aircraft	Engine	Remarks
Albatros D.XI	Siemens-Halske Sh.III	
Albatros D.XII	B.M.W.IIIa	
Fokker V.28	Siemens-Halske Sh.III	Monoplane, prototype of Fokker E.V
Fokker V.29	B.M.W.IIIa	Co-winner of competition, monoplane
Fokker V.36	B.M.W.IIIa	Modified Fokker D.VII
Fokker D.VIII	Oberursel Ur.III	Production monoplane with bigger engine
Junkers D.I	B.M.W.IIIa	Monoplane, in small-scale production for evaluation
Kondor E.III	Oberursel Ur.III	Monoplane, small evaluation batch produced
Kondor E.IIIa	Goebel Goe.III	Monoplane, small evaluation batch produced
Pfalz D.XVf	B.M.W.IIIa	Placed in production
Pfalz D.XV (special)	B.M.W.IIIa	
Roland D.XVI	Siemens-Halske Sh.III	Monoplane
Roland D.XVII	B.M.W.IIIa	Monoplane
Rumpler D.I	B.M.W.IIIa	Co-winner of competition, biplane
Zeppelin D.I	B.M.W.IIIa	Evaluation batch ordered

Notes:

1. Two aircraft were selected as winners, the Fokker V.29 and Rumpler D.I. The Fokker V29, a larger, heavier, and more powerful aircraft than the rotary-powered E.V/D.VIII, shared its basic wing design with the E.V/D.VIII. Given the fatal wing failures of the Fokker E.V and the continued superiority of the Fokker D.VII at the front, *Idflieg* had not ordered the V29 into production by the armistice.
2. Monoplane competitors out-numbered biplanes eight to seven, a marked reversal of the 1917 triplane fetish.
3. The Pfalz D.XV had been ordered into production before the competition to replace the Pfalz D.XII. The Pfalz company, located in Bavaria, was supported by Bavarian authorities regardless of the competitions.

Below: The Zeppelin D.I fighter featured an all-metal structure but had fabric covering over the horizontal tail surfaces and on the wings behind the metal box spars. The upper wing attachment structure had been significantly strengthened since the fatal accident at the Second Fighter Competition. D.I 1751/18 (Ersatz) shown here was flown at this competition.

Above & Below: Aces at the Third Fighter Competition to evaluable the prototype fighters. Left to right, *Lt.* Walter Blume, *Lt.* Josef Veltjens, *Lt.* Josef Jacobs, *Oblt.* Oskar von Boenigk, *Hptm.* Eduard *Ritter* von Schleich, *Lt.* Ernst Udet, *Hptm.* Bruno Loerzer, *Lt.* Paul Baümer, *Oblt.* Hermann Göring, *Lt.* Heinrich Bongartz.

Above: At the Third Fighter Competition; left to right, *Hptm von* Schleich, *Hptm* Lorezer, *Oblt*. Göring, *Lt*. Udet, *Lt*. Bongartz.
Bottom: At the Third Fighter Competition; left to right, *Lt*. Udet, *Hptm*. Loerzer, and an *Idflieg* officer.

Fokker at the Third Fighter Competition

After winning the first two fighter competitions, Fokker entered the Third Fighter Competition in the best position of any manufacturer. The Fokker D.VII was the mainstay of the German fighter units and the BMW-powered version was generally acknowledged as the best all-around fighter in combat. Fokker's D.VII was being built not only by Fokker but by other manufacturers under license. After a rocky start due to two fatal wing failures, Fokker D.VIII production was back in full swing.

Above: Fokker D.VII mass production at Fokker.

Fokker's entries in this competition built on these successes. The V28 was a D.VIII prototype with the more powerful Seimens-Halske rotary, the V36 was a modified D.VII, and the V29 was a parasol monoplane derivative of the D.VII. Using the same basic configuration as the production D.VIII, the V29 depended on its over-compressed BMW engine for excellent high-altitude performance and was one of the winners of the competition. However, with fresh memories of the D.VIII wing debacle and with the robust, proven D.VII still superior, *Idflieg* was in no hurry to order the V29 into production.

Fokker V29 Specifications

Engine:	185 hp BMW.IIIa	
Wing:	Span	9.68 m
	Area	14.3 m^2
General:	Length	7.02 m
	Height	2.93 m
	Empty Weight	632 kg
	Loaded Weight	861 kg
Maximum Speed:		200 km/h
Climb:	3000m	7.5 min
	4000m	10.5 min
	5000m	14.7 min
	6000m	19.7 min

Below: The Fokker V29 shown here was one of the two winners of the Third Fighter Competition. Essentially, it was a parasol monoplane derivative of the Fokker D.VII. Similar to the smaller, D.VIII, it had the more powerful 185 hp BMW.IIIa engine for better high altitude performance. Similarly, the more powerful D.VII biplane had eclipsed the smaller, rotary-powered D.VI of similar configuration.

Above: The Fokker D.VIII was in mass production at Fokker and had been re-introduced to service with combat units by the end of the Third Fighter Competition.

Below: The Fokker V36 was derived from the production D.VII. Although it had a modified radiator, its primary innovation was housing the main fuel tank in the enlarged airfoil around the undercarriage axle, reducing the hazard in case of an in-flight fire due to combat damage.

Fokker V36 Specifications

Engine:	185 hp BMW.IIIa	
Wing:	Span	8.935 m
	Area	17.6 m^2
General:	Length	6.46 m
	Height	3.045 m
	Empty Weight	637 kg
	Loaded Weight	871 kg
Maximum Speed:		200 km/h
Climb:	1000m	1.8 min
	2000m	4.0 min
	3000m	6.8 min
	4000m	10.0 min
	5000m	13.5 min
	6000m	18.3 min

This page: The Fokker V36, derived from the production D.VII, had a modified radiator, but its primary innovation was housing the main fuel tank in the enlarged airfoil around the undercarriage axle, reducing the fire hazard in case of an in-flight fire due to combat damage.

Rumpler at the Third Fighter Competition

Rumpler's situation at the Third Fighter Competition continued to be frustrating. Rumpler had been working on its high-altitude D.I fighter since early 1916, and prototypes had given good performance at the first two fighter competitions, but despite being declared one of the winners of the Third Fighter Competition, the D.I had not yet passed its type test started in February and was *still* not ready for front-line service.

Despite a stronger fuselage, the D.I wings still vibrated in flight, it did not glide well, and it did not make its pilots feel secure. In particular, aileron response was slow and erratic due to the sophisticated but unproven worm-gear actuation system, a serious drawback in a fighter. Furthermore, the short fuselage, again chosen to minimize weight, gave limited pitch stability and made landings difficult. During a review prior to the Third Fighter Competition *Idflieg* documented 42 installation faults. Although most were minor, the sheer number of descrepancies calls into question the competence of the engineering team and their ability to deliver a combat-worthy fighter.

Rumpler D.I Specifications

Engine:	185 hp BMW.IIIa	
Wing:	Span Upper	8.42 m
	Span Lower	7.38 m
	Chord Upper	1.32 m
	Chord Lower	1.09 m
	Gap	1.365 m
	Area	16.66 m^2
General:	Length	5.75 m
	Height	2.56 m
	Empty Weight	615 kg
	Loaded Weight	805 kg
Maximum Speed (at 5000 m):		180 km/h
Climb:	5000m	13.2 min
	6000m	17.8 min
	7000m	26.9 min
	8000m	53.3 min
Ceiling:		8200m

Despite excellent climb and ceiling and now fairly good flight characteristics, the Rumpler D.I was judged of limited use in fighter versus fighter combat and was *still* not ready for production! The Daimler L11 was flying and the SSW D.VI was in progress…

Above: The other winner of the Third Fighter Competition was the Rumpler D.I, which offered exceptional ceiling and high-altutide performance. Rumpler struggled with its prolonged development and it never gelled into a production fighter. D.I 1581/18 used the 185 hp BMW.IIIa engine to compete at the Third Fighter Competition, the only D.I to do so.

Pfalz at the Third Fighter Competition

The Pfalz D.XV was the ultimate Pfalz fighter design. The new design was clearly evolved from the D.XII, yet was strongly influenced by the exceptional Fokker D.VII. The D.XV wings were thicker than pervious Pfalz designs, although not as thick as Fokker D.VII wings, and featured single bays. Early prototypes had minimum wire bracing, but the design was finally developed through a number of prototypes to a wireless wing like the Fokker D.VII.

The production Pfalz D.XV was a fast, maneuverable fighter that the Germans considered the equal of the Fokker D.VII, which was high praise considering the Fokker D.VII was probably the best operational fighter of WWI. Powered, like the best D.VIIs, by the 185 hp BMW.IIIa engine, the D.XV demonstrated an excellent climb rate. Production started in October 1918, and the *Typenprüfung* was formally completed on November 4, 1918, a week before the Great War ended. Apparently some production aircraft reached supply depots (*Flugparks*) before the war ended, but had not yet been delivered to operational units.

The Austro-Hungarian *Luftfahrtruppe* was also interested in the Pfalz D.XV and planned license production; these plans were cancelled with the armistice. When the war ended, so did Pfalz's chance to finally step from the shadow of Fokker and make a solid reputation for itself with the excellent Pfalz D.XV.

Pfalz D.XV Specifications

Engine:	185 hp BMW.IIIa	
Wing:	Span Upper	8.60 m
	Span Lower	7.20 m
	Chord Upper	1.45 m
	Chord Lower	1.20 m
	Stagger	0.36 m
	Area	20 sq m
General:	Length	6.50 m
	Height	2.70 m
	Empty Weight	738 kg
	Loaded Weight	928 kg
Maximum Speed:		200 kmh
Climb:	1000m	2 min
	2000m	5 min
	3000m	8 min
	4000m	11.5 min
	5000m	16 min
	6000m	22.5 min

Left: Pfalz 'experimental D.XV No.1' was powered by the 180 hp Mercedes D.IIIa. The fin is integrated with the fuselage and the lower wing is attached via a keel. The I-struts restricted view during combat.

Below: Pfalz 'experimental D.XV No.2' with N-struts, the lower wing supported by struts, and the fin no longer integral with the fuselage.

Above & Below: Pfalz 'experimental D.XV No.1' was powered by the 180 hp Mercedes D.IIIa. The fin is integrated with the fuselage and the lower wing is attached via a keel. The I-struts reduced drag but restricted view during combat.

Above: The Pfalz D.XV, finally with the 185 hp BMW.IIIa engine the D.XII had needed to reach its full potential, was an excellent fighter developed from the D.XII. It was as fast and maneuverable as the Fokker D.VII, and the first examples arrived at the Army *Flugparks* the last week of the war. Missing combat, it was forgotten in the aftermath. This photo illustrates the production version with horn-balanced ailerons.

Above: Two Pfalz D.XV fighters, both powered by the 185 hp BMW.IIIa engine, were entered in the Third Fighter Competition. Although not declared winners, the semi-autonomous Bavarian authorities had already ordered it into production to replace the Pfalz D.XII from which it was developed. This late prototype may have been the 'special' D.XV entered in the competition; it lacked the horn-balanced control surfaces used on the production D.XV.

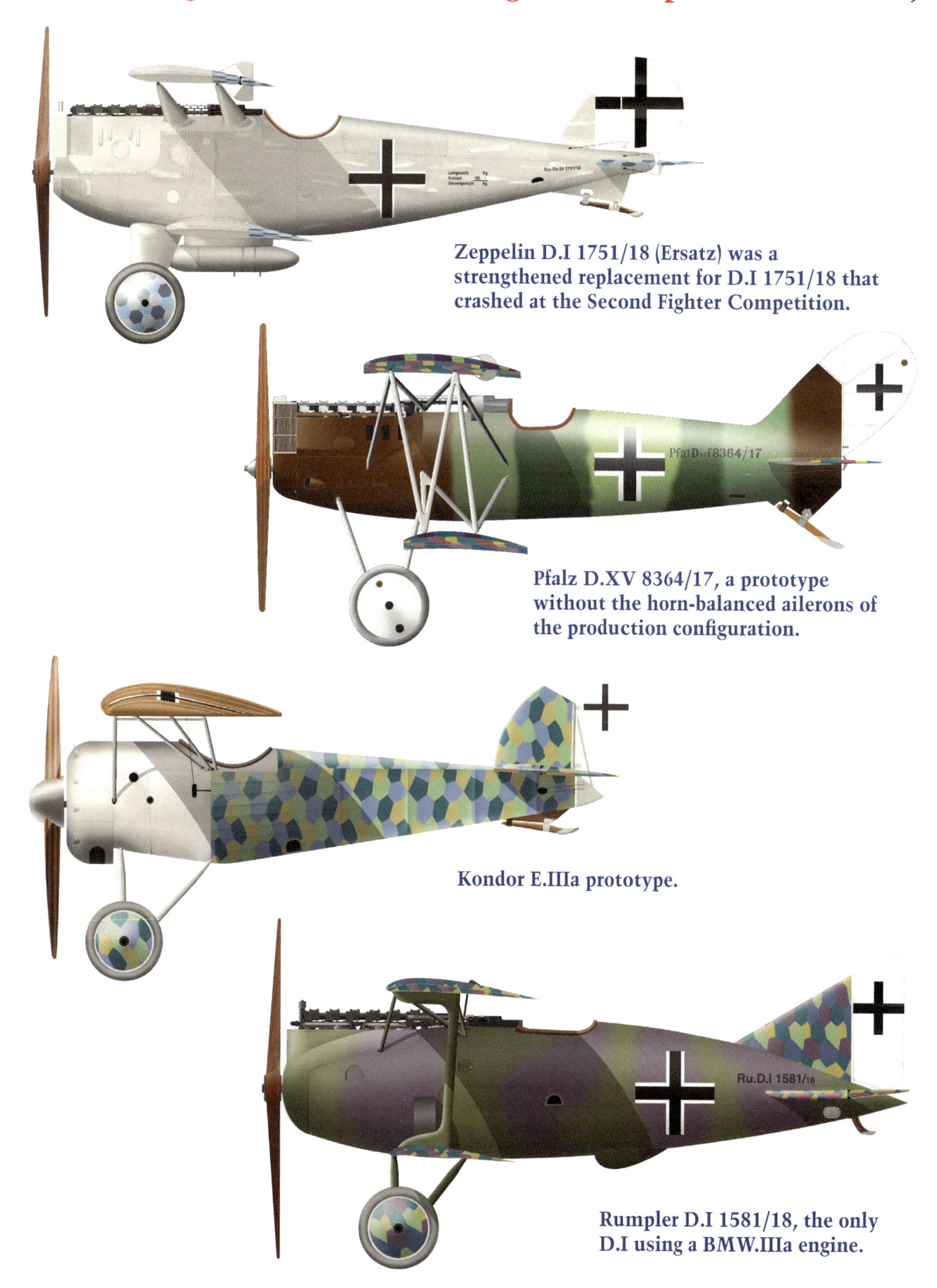

Zeppelin D.I 1751/18 (Ersatz) was a strengthened replacement for D.I 1751/18 that crashed at the Second Fighter Competition.

Pfalz D.XV 8364/17, a prototype without the horn-balanced ailerons of the production configuration.

Kondor E.IIIa prototype.

Rumpler D.I 1581/18, the only D.I using a BMW.IIIa engine.

Albatros at the Third Fighter Competition

The first two fighter competitions had been major disappointments to Albatros. Albatros had lost its fighter dominance to rival Fokker, and was forced to build the Fokker D.VII under license. Worse yet, the Albatros fighters had fallen behind not just Fokker but a number of other manufacturers.

If the results of the Second Fighter Competition were disppointing to Albatros, the Third Fighter Competition had to be especially frustrating. The redesigned Albatros D.XI, 2208/18, now landed more easily due to the shorter undercarriage made possible for its new four-bladed propeller, and flying characteristics were good, but the roll rate, so important to a fighter, was too slow. It was included in the competition as 'an interesting type' despite the official limitation to BMW-powered fighters.

The Albatros D.XII, which had been too late to compete officially in the Second Fighter Competition but was in time for evaluation by the front-line pilots, had now been improved by further development. The revised D.XII had a longer fuselage and wing-span and aerodynamically-balanced ailerons.

Surprisingly, the flight characteristics of the revised Albatros D.XII were now judged as superior to those of the Fokker D.VII. However, the D.XII was slower than the Rumpler D.I and slightly inferior to the Fokker V29 and V36. Moreover, despite extensive wire bracing, the airframe was not as robust as the competing Fokker D.VII and V36. With the sturdy D.VII in full production and widespread and successful service, and also being faster than the D.XII, Albatros again was frustrated in securing a production order for its new fighter.

Albatros D.XII Specifications

Engine:	185 hp BMW.IIIa	
Wing:	Span Upper	8.85 m
	Span Lower	7.42 m
	Chord Upper	1.40 m
	Chord Lower	1.40 m
	Area	20.01 m^2
General:	Length	5.78 m
	Height	2.80 m
	Empty Weight	621 kg
	Loaded Weight	801 kg
Maximum Speed		180 kmh
Climb:	1000m	1.9 min
	2000m	4.1 min
	3000m	6.8 min
	4000m	10.4 min
	5000m	16.7 min

Above: Considered an 'interesting type' Albatros D.XI 2208/18 was included in the Third Fighter Competition despite being powered by a Siemens-Halske rotary. The shorter undercarriage legs allowed by the four-blade propeller made this aircraft easier to land than Albatros D.XI 2209/18 that competed in the Second Fighter Competition. The D.XI prototypes were the only Albatros fighters powered by rotary engines. Struts braced the wings, resulting in a 'wireless' design. However, despite improved landing and flying qualities and competitive performance, roll rate was too slow, a critical liability in a fighter, and no production was ordered.

Above: The Albatros D.XII 2210/18 powered by the Mercedes D.III was evaluated in July, where it was judged slightly inferior to the Pfalz D.XII and was not chosen for production. The modified D.XII 2210/18 shown above was now powered by the 185 hp BMW.IIIa and competed at the Third Fighter Competition. The D.XII was the last Albatros fighter design flown during the war and the last chance for Albatros to regain its former place in fighter design.

Above: Albatros D.XII 2211/18 powered by the Mercedes D.III featured a longer fuselage, modified wing, and aerodynamically-balanced ailerons. Tests were still in progress on this airframe as late as September 1918 as Albatros's final attempt at a replacement for the Albatros D.Va. The D.XII now had good flying qualities but was not robust enough despite its extensive wire bracing, which created too much drag and made it slower than other competitors.

Above & Below: Considered an 'interesting type' Albatros D.XI 2208/18 was included in the Third Fighter Competition despite being powered by a Siemens-Halske rotary. The shorter undercarriage legs allowed by the four-blade propeller made this aircraft easier to land than Albatros D.XI 2209/18. The D.XI prototypes were very compact, but the roll rate was too low for a successful fighter, likely due to the close proximity of the wings spars needed for the I-struts.

Roland at the Third Fighter Competition

Roland had been frustrated at the Second Fighter Competition and did no better at the Third Fighter Competition despite entering two new parasol monoplane fighters. Roland's wing design was inferior to Fokker's wing; the rotary D.XVI offered no advantage over the SSW D.IV and Fokker D.VII, and the D.XVII was markedly inferior to the Fokker V29, exhibiting alarming wing vibration when subjected to maneuvering loads.

Roland D.XVI Specifications		
Engine:	160 hp Siemens-Halske Sh.III	
Wing:	Span	9.46 m
General:	Length	5.90 m

Roland D.XVII Specifications	
Engine:	185 hp BMW.IIIa

Above: The Roland D.XVI competed at the Third Fighter Competition despite being powered by the 160 hp Sh.III rotary. During most of 1917 *Idflieg* encouraged designers to try triplanes, but during 1918 monoplanes became increasingly popular for fighter designs due to their greater speed, and the last two Roland fighter prototypes were monoplanes.

Above: The Roland D.XVII fighter powered by the 185 hp BMW.IIIa competed at the Third Fighter Competition.

Zeppelin at the Third Fighter Competition

Zeppelin D.I D.1751/18 (*Ersatz*) was built to replace the aircraft that Reinhard crashed. This new D.I was powered by the 185 hp BMW.IIIa engine and participated in the Third Fighter Competition. The BMW engine substantially improved the rate of climb compared to the Mercedes D.III.

Unfortunately, the Zeppelin D.I overturned on landing on 25 October, the day before the combat pilots started to evaluate the competitors. Regardless, Junkers company test pilot *Lt.* Krohn reported that at the final debriefing the combat pilots stated their preferences, in order, as:

1. Junkers D.I (J9)
2. Rumpler D.I
3. Albatros D.XII
4. Zeppelin D.I
5. Fokker D.VII (V36)

Krohn's comments are not entirely consistent with the fact that the Fokker V29 and Rumpler D.I were declared the winners of the Third Fighter Competiton and that the combat pilots generally did not like the low-wing configuration of the Junkers D.I that Krohn championed. Regardless, *Idflieg* ordered a batch of 50 Zeppelin D.I fighters, 1900–1949/18, but production was halted in early 1919 and only a few aircraft were completed. After the war the US Air Service purchased one, either D.1754/81 or D.1755/18, and evaluated it as AS-68546. Similarly, the US Navy evaluated another D.I, D.1753.18. The final prototype was exhibited in the Dornier Museum in Friedrichshafen but was destroyed by Allied bombing during WW2.

Zeppelin D.I Specifications

Engine:	185 hp BMW.IIIa	
Wing:	Span	7.80 m
	Area	16.66 m²
General:	Length	6.40 m
	Height	2.60 m
	Empty Weight	725 kg
	Loaded Weight	883 kg
Maximum Speed:		200 km/h
Climb:	1000m	2.6 min
	5000m	13 min
	7600 m	51 min

Above: Zeppelin D.I 1751/18 (Ersatz) is seen here at the Third Fighter Competition. Powered by the 185 hp BMW.IIIa, a similar aircraft was evaluated by the US Air Service at McCook Field in 1922. The aircraft is in overall natural finish, with the fabric in three-color German naval camouflage fabric (Zeppelin also built some all-metal Giant flying boats for the German Navy and apparently had extra fabric). The fuel tank was under the fuselage to minimize danger from fire, and reportedly was planned to be jettisonable.

Kondor at the Third Fighter Competition

Newcomer Kondor had not been a factor at the first two fighter competitions, but made a surprise hit at the Third Fighter Competition with its E.III and E.IIIa parasol monoplane fighters. While established companies like Albatros and Roland were fizzling out, Kondor designer Rethel was inspired by the Fokker E.V and invented a new wing structure that was very strong yet light weight. The resulting Kondor E.III had veneer sheets applied chordwise between ribs that projected above the wing surface. The protruding ribs were claimed to offer aerodynamic benefits, and may have improved boundary layer flow.

Flight evaluation showed the E.III to have much better flight characteristics than the Fokker D.VIII and its wing was much stronger. *Hptm*. Eduard *Ritter* von Schleich, commander of *JG*4, extolled the E.III as the best fighter at the competition and claimed to have arranged for an E.III to be shipped

Kondor E.III Specifications

Engine:	160 hp Oberursel Ur.III	
Wing:	Span	9.00 m
	Wing Area	12.75 m^2
General:	Length	5.50 m
	Height	2.75 m
	Empty Weight	460 kg
	Loaded Weight	640 kg
Maximum Speed:		190 km/h
Climb:	5000m	16.0 min
Ceiling:		6,180 m

Kondor E.IIIa Specifications

Engine:	160 hp Goebel Goe.III	
Wing:	Span	9.00 m
	Wing Area	12.75 m^2
General:	Length	5.50 m
	Height	2.75 m
Maximum Speed:		200 km/h
Climb:	5000m	11.0 min

Above & Below: The Kondor E.IIIa was powered by the 160 hp Goebel Goe.III rotary; the E.IIIa had a spinner and full cowl.

directly to him. The Kondor E.III was loaded on a flatcar on November 2 but never arrived due to the Armistice on the 11th. Kondor claimed 100 fighters, now officially designated the Kondor D.I (previous Kondor fighter designations had been factory, not official designations), were ordered but there is no confirming documentation.

The Kondor E.III used the 160 hp Oberursel Ur.III, and the E.IIIa was powered by the 160 hp Goebel Goe.III. Due to the different engine, the E.IIIa, which was faster and had a much better climb rate, used a different cowling than the E.III and had a spinner.

Apparently a small number of E.III/E.IIIa aircraft, probably no more than ten, were built, and several were sent to *Idflieg* for flight testing. But it was now too late; the war was over and Kondor's triumph remained virtually unknown. A few found their way to the civilian market after the war.

Above: The Kondor E.III was powered by the 160 hp Oberursel Ur.III rotary. The E.III had a cutaway cowl and no spinner was fitted. No armament is fitted. This E.III competed in the Third Fighter Competition.

Below: The Kondor E.IIIa was powered by the 160 hp Goebel Goe.III rotary; the E.IIIa had a spinner and full cowl. This aircraft was photographed in 1919 in Switzerland, where it was employed for aerobatic displays.

Above: This Kondor E.IIIa was photographed in 1919 in Switzerland, it was owned by the Comte, Mittleholzer & Co. (Aero Gesellschaft). An LVG C.V is at right.

Below: A Kondor E.IIIa at Cuyk, Holland, in 1920; the pilot is Hans Wende of NAVO. The unusual wing structure is evident. Unlike the Fokker D.VIII, the cantilever wing did not vibrate at high speed. The stronger wing structure of the E.III/IIIa also gave it a better roll rate than the Fokker D.VIII, which is very important for a fighter.

Junkers at the Third Fighter Competition

Development of the Junkers D.I single-seat fighter contined between the Second and Third Fighter Competitions due to *Idflieg's* support, and the type-test aircraft, 5180/18, was approved for front-line service on 9 September 1918 after final type testing.

For the Third Fighter Competition Junkers entered a D.I fitted with BMW engine, longer fuselage, and increased wingspan. The enlarged dimensions reduced maneuverability and performance. However, the Junkers D.I still bested all but one of the competitors in rate of climb and was faster than the Fokker V36, although the Fokker V29 was faster above 4,000 meters.

Now joined by the Zeppelin D.I, the Junkers D.I was no longer the world's only all-metal fighter, but it was already in production as the world's first all-metal *production* fighter. Several production aircraft reached Belgium before the end of the war, but nothing is known of their combat use, if any.

Junkers D.I Specifications		
Engine:	185 hp BMW.IIIa	
Wing:	Span	9.00 m
	Area	14.8 m^2
General:	Length	6.70 m
	Length (late version)	7.25 m
	Height	2.60 m
	Empty Weight	654 kg
	Loaded Weight	834 kg
Maximum Speed:		225 km/h
Climb:	5000m	24.0 min

Howerver, one airframe had been struck 'by several bursts of machine gun bullets,' indicating combat. Postwar the D.I was used very successfully in combat in severe winter weather and stood up to the elements far better than wooden aircraft. According to *Lt.* Sachsenberg who flew them, "The Junkers have proven themselves beyond all expectation."

Above: Although not a winner of the Third Fighter Competition, the Junkers D.I had already been placed in production to become the world's first production all-metal fighter. Examples arrived in Belgium just before the Armistice. This aircraft has a 185 hp BMW.IIIa engine for the Third Fighter Competition and has the lengthened fuselage and larger wingspan.

The German Fighter Force in 1919

Above: The SSW D.VI was not completed until after the armistice; by then the 'E' category for monoplane fighters was no longer used. Powered by the Sh.IIIa high-altitude counter-rotary engine then giving 220 hp, it had a top speed of 220 kmh (137 mph) coupled with excellent maneuverability and climb and a ceiling of 8,000m. The under-fuselage fuel tank was jettisonable in case of fire. The SSW D.VI was the natural production successor to the SSW D.IV.

It is interesting to speculate on what the front-line German fighter force would have looked like had the war continued into 1919. In fact, we know enough about German fighters to have a fairly good idea for at least the first half of 1919.

Carry Over Designs from 1918

Of course, most of the front-line inventory of late 1918 would have carried over into 1919. By early 1919 most of the obsolete Albatros, Roland, and Pfalz D.III/D.IIIa fighters and Fokker Triplanes, and perhaps all, would have been relegated to the flight schools and would have disappeared from the front.

The excellent Fokker D.VII, still unsurpassed by Allied fighters, was in wide-spread service and production in late 1918 and initially would have been the largest component of the fighter inventory. The Fokker D.VIII would also have been in service, perhaps along with limited numbers of the Fokker D.VI at the second-line *Jastas*.

The Pfalz D.XII, while relatively numerous at the front, was unloved by pilots and likely would have been very quickly replaced in the *Jastas* by its far superior development, the Pfalz D.XV. In fact, the D.XV reached the *Armee Flugparks* in the last week of the war but was too late to see combat.

The SSW D.III and D.IV were in service in small numbers and would have lasted into 1919 until eliminated by attrition. The similar Pfalz D.VIII was also serving in small numbers and likewise would have served in small numbers into 1919.

Like the Pfalz D.XV, the all-metal Junkers D.I had just reached the front before the armistice. A fast, robust fighter, it would have seen service in greater numbers in 1919, but the difficulty of producing the Junkers fighter would have limited its numbers.

New Fighters at the Front

The two winners of the Third Fighter Competition, the Fokker V29 and the Rumpler D.I, may have been cautiously introduced into service in small numbers.

The painful memories of the Fokker D.VIII wing failure debacle were certainly fresh, and the Fokker V29 had a very similar configuration. The rotary-powered D.VIII had sufficient wing strength when properly manufactured, but the V29 was heavier and faster, so it would have needed careful testing and quality control during production to ensure its reliability in combat. Furthermore, when it went into production it would have replaced the Fokker D.VII, still the best fighter at the front and proven to be robust and reliable. *Idflieg* was in no hurry to replace the proven D.VII in production as long as it was superior to its adversaries.

The Rumpler D.I was also very problematic. Development of this aircraft had been underway

Above: *Jasta* 40 Fokker D.VIIs. At the Armistice the Fokker D.VII with BMW.IIIa retained its superiority over Allied fighters and was serving in the *Jastas* in much greater numbers than any other fighter. There is no doubt it would have represented the bulk of the German fighter force in early 1919 had the war continued.

Above: *Oblt.z.S.* Gotthard Sachsenberg's spectacular black and yellow Fokker D.VII; Sachsenberg scored 31 victories.

Above: Unknown *Jasta* 28 pilot in front of his Fokker D.VII.

Above: Fokker D.VII 6810/18 on display in Canada wears it original camouflage. When fitted with the over-compressed 185 hp BMW.IIIa the Fokker D.VII established superiority over Allied fighters for the rest of the war.

Above: The Fokker V29 parasol monoplane derivative of the Fokker D.VII was one of the two winners of the Third Fighter Competition and possibly would have been produced as a successor to the Fokker D.VII. However, *Idflieg* was cautious about the reliability of its cantilever monoplane wing given the recent debacle with similar wings on the Fokker E.V, plus the robust, proven D.VII was still superior to Allied fighters at the front, so no orders were placed before the Armistice.

Right: The Fokker V29 at the Third Fighter Competition.

Fokker V29 prototype, co-winner of the Third Fighter Competition.

since early 1916 and bringing it to practical production status was proving very difficult. Even the version flown in the Third Fighter Competition was not fully ready for production, and there is therefore a serious question whether this warplane could have achieved practical production status before being matched or exceeded in performance by other, more robust and practical designs.

Interestingly, the Pfalz D.VII, initially evaluated at the First Fighter Competition, would likely have been seen in combat in 1919. About 30 Pfalz D.VII fighters were found stored after the war, indicating that Pfalz had solved its wing cellule strength limitations. The single-bay D.VII was lighter and had less drag than the two-bay D.VIII, giving the D.VII better speed and climb rate, and the D.VII would likely have replaced the D.VIII for that reason.

It is very likely the SSW D.III and D.IV biplanes would have been replaced in production and front-line service by the advanced SSW D.VI parasol monoplane. The SSW D.VI was 30 km/h faster than its D.IV predecessor due to its lower drag and use of a more powerful version of the Siemens-Halske D.IIIa engine that powered the D.IV, and climb rate and ceiling were nearly identical. In fact, the SSW D.VI offered the climb and ceiling the Rumpler D.I promised coupled with greater speed and maneuverability, plus technology already proven in the earlier D.III and D.IV, perhaps making the SSW D.VI the fighter that reached the front instead of the troublesome Rumpler D.I.

Finally, there likely would have been some surprises. One would have been the Kondor E.III/IIIa parasol monoplane, military designation D.I, that was so impressive when evaluated. Another would have been the Aviatik D.VII biplane. Despite not being entered in the Third Fighter Competition, perhaps because it used the new Benz Bz.IIIbm V-8 engine and the Third Fighter Competition was restricted to aircraft using the BMW.IIIa, the Aviatik D.VII was actually placed in production. During its post-war inspections the Inter-Allied Control Commission found 50 production Aviatik D.VII fighters hidden away in storage with no mention of their existence. Had the war continued into 1919, these fighters, the first production German fighters powered by a V-8 engine, would have arrived at the front, perhaps reviving a manufacturer's name famous early in the war but little heard from since.

The Zeppelin D.I was already in limited production and would likely have been evaluated in small numbers at the front. Finally, the virtually unknown Daimler L 11, powered by Daimler's 185 hp Mercedes D.IIIbm geared V-8 and not at the Third Fighter Competition, had exceptional speed and climb and may well have entered service as the Daimler D.III, rendering the intractable Rumpler D.I unnecessary.

Above: Apparently Pfalz was able to strengthen the single-bay wing of their D.VII sufficiently to put it into production, because about 30 production D.VII fighters were found in Germany after the war by the Inter-Allied Control Commission. The D.VII was lighter and had less drag than its two-bay D.VIII sibling and would have been a natural to replace it in modest production as an interceptor. This D.VII has experimental horn-balanced ailerons and was powered by a 145 hp Oberursel Ur.III rotary, but production versions had the 160 hp Siemens-Halske Sh.III rotary and no horn balances.

Above: The all-metal Junkers D.I had just started to arrive at the front before the Armistice and its numbers would have built slowly into 1919 as production gained momentum. However, the comparatively slow production rate of this new technology aircraft would have prevented it from becoming a large portion of the German fighter force.

Above & Right: Fokker E.V fighters briefly in service with *Jasta* 6 in early August 1918. After fatal crashes due to wing failures, the E.V was removed from the front and all wings were replaced. The rebuilt aircraft returned to the front in October, now designated the Fokker D.VIII. Had the war continued the Fokker D.VIII would have remained at the front into 1919.

Above & Below: In fairly widespread service in late 1918, the unloved Pfalz D.XII (above preserved at the Australian War Memorial) would have lasted at the front into 1919. However, during early 1919 it would have been rapidly replaced by the production Pfalz D.XV (below), derived from the D.XII but a faster, much more maneuverable fighter powered by the over-compressed 185 hp BMW.IIIa. The Germans considered the Pfalz D.XV equivalent to the Fokker D.VII.

This Page: Three views of an armed but unmarked D.XV photographed in May 1919 at Villacoublay while being evaluated by the French. Pfalz managed to eliminate drag-producing bracing wires in the D.XV, which also eliminated a lot of airframe maintenance. The D.XV also offered excellent handling and maneuverability, a breakthrough for Pfalz.

Left & Below: The SSW D.IV (left) and D.III (below) served in small numbers in 1918, limited primarily by engine availability. They would have carried over into 1919 until replaced in production and at the front by the significantly faster SSW D.VI parasol monoplane.

Below: The two-bay Pfalz D.VIII served in small numbers during 1918. In 1919 it would have been gradually replaced by the slightly faster single-bay Pfalz D.VII that had gone into production when structural modifications improved its wing strength.

This Page: The SSW D.VI parasol monoplane demonstrated excellent speed, climb, and ceiling and there is little doubt it would have replaced the SSW D.IV biplane in production in early 1919 had the war continued. Continued engine development may have enabled it to serve reliably as a general-purpose fighter as well as an interceptor, perhaps filling the role intended for the troubled Rumpler D.I.

SSW D.VI Specifications

Engine:	220 hp Siemens-Halske Sh.IIIa	
Wing:	Span	9.37 m
	Area	12.46 m^2
General:	Length	6.50 m
	Height	2.700 m
	Empty Weight	540 kg
	Loaded Weight	710 kg
Maximum Speed:		220 km/h
Climb:	6000m	16 min
	7000m	22 min
Ceiling:		8000m

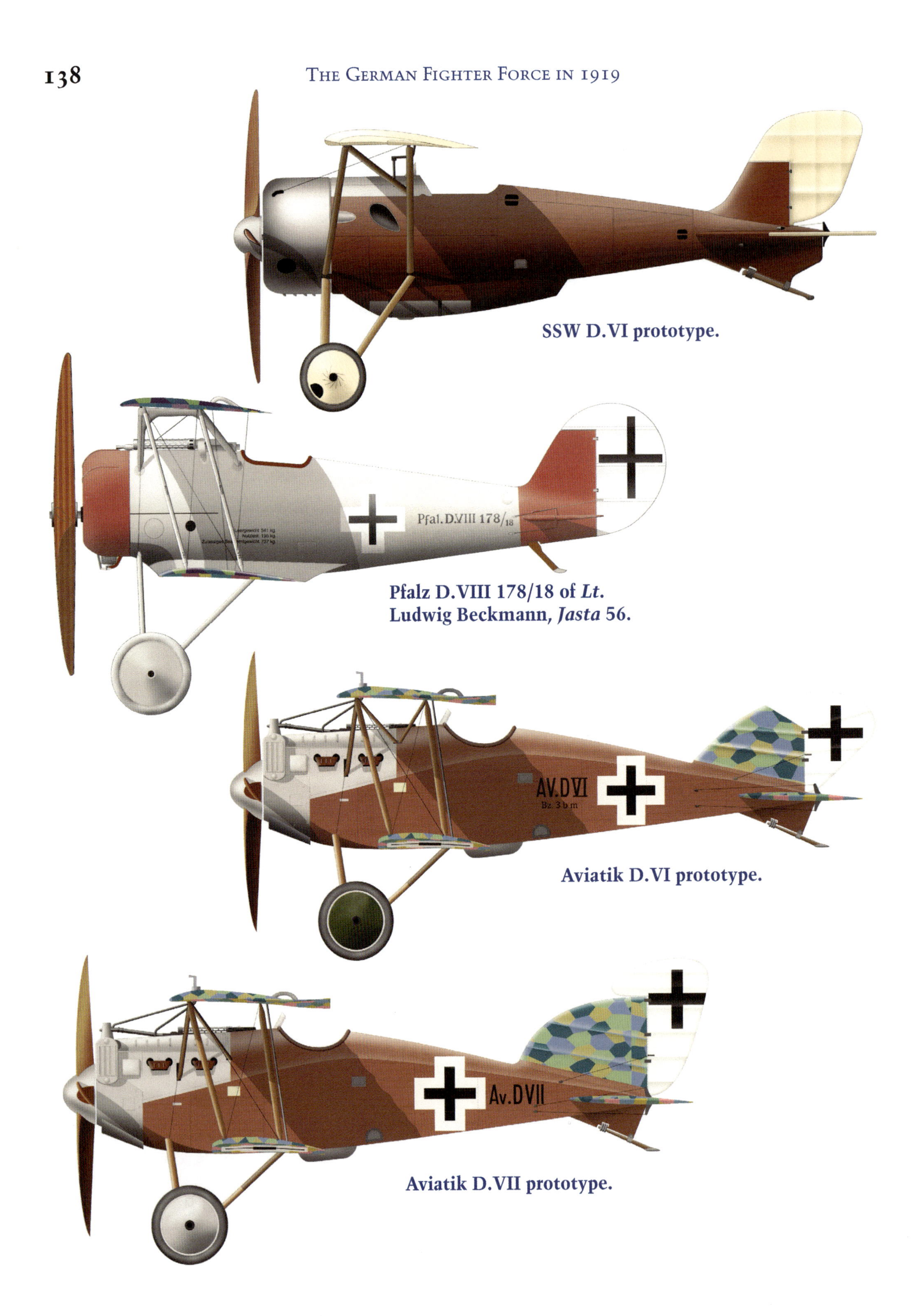

SSW D.VI prototype.

Pfalz D.VIII 178/18 of *Lt.* Ludwig Beckmann, *Jasta* 56.

Aviatik D.VI prototype.

Aviatik D.VII prototype.

Above: The Rumpler D.I, subject to a long and painful development process, was entered in all three Fighter Competitions and was one of the two winners of the Third Fighter Competition. It may have been produced in 1919 for the German Air Service as a high-altitude fighter if Rumpler could ever manage to get it to meet acceptance standards.

Below: This closeup of the SSW D.VI shows the exhaust outlet and the jettisonable belly fuel tank to advantage.

Aviatik D.VII

Aviatik D.VII Specifications		
Engine:	195 hp Benz Bz.IIIbm V-8	
Wing:	Span	9.66 m
General:	Length	6.10 m
	Height	2.50 m
	Empty Weight	745 kg
	Loaded Weight	945 kg
Maximum Speed:		192 kmh
Climb:	6000m	24.0 min

These three photos are of the largely unknown Aviatik D.VII fighter. A development of the Aviatik D.VI that was evaluated in conjunction with the Second Fighter Competition, the Aviatik D.VII had substantially revised tail surfaces but was otherwise essentially the same as the D.VI and was powered by the same Benz Bz.IIIbm V-8 used by the D.VI.

The Third Fighter Competition was for aircraft powered by the BMW.IIIa engine, so the Aviatik D.VII did not participate. Nevertheless, it was covertly placed in production and the Inter-Allied Control Commission found 50 completed Aviatik D.VII fighters after the war.

Above: Powered by a 160 hp Goebel Goe.III rotary, the Kondor D.IIIa competed at the Third Fighter Competition. This is a 1919 view of an unarmed Kondor D.IIIa in civil Swiss service with the owner, Alfred Comte, in the cockpit.

Below: The all-metal Zeppelin D.I fighter had the most advanced structure of any aircraft at the Third Fighter Competition. Prototypes were powered by both the Mercedes D.IIIa and by the 185 hp BMW.IIIa. The surface of the wings aft of the box spar had fabric covering, as did the horizontal tail surfaces. This example was brought to the USA for testing after the war and is shown after re-assembly. Other than its biplane configuration, this was the preview of future propeller-driven fighters until the jet age; then early jets used similar structural technology. Postwar the Zeppelin D.I has often been referred to as the 'Dornier D.I' in honor of its brilliant designer, *Dipl-Ing.* Claude Dornier, who later founded his own company, but Zeppelin D.I, the company name during the war (or Zeppelin-Lindau D.I after the division of the company), is the more correct designation because that was what it was called when it was designed, built, and flown.

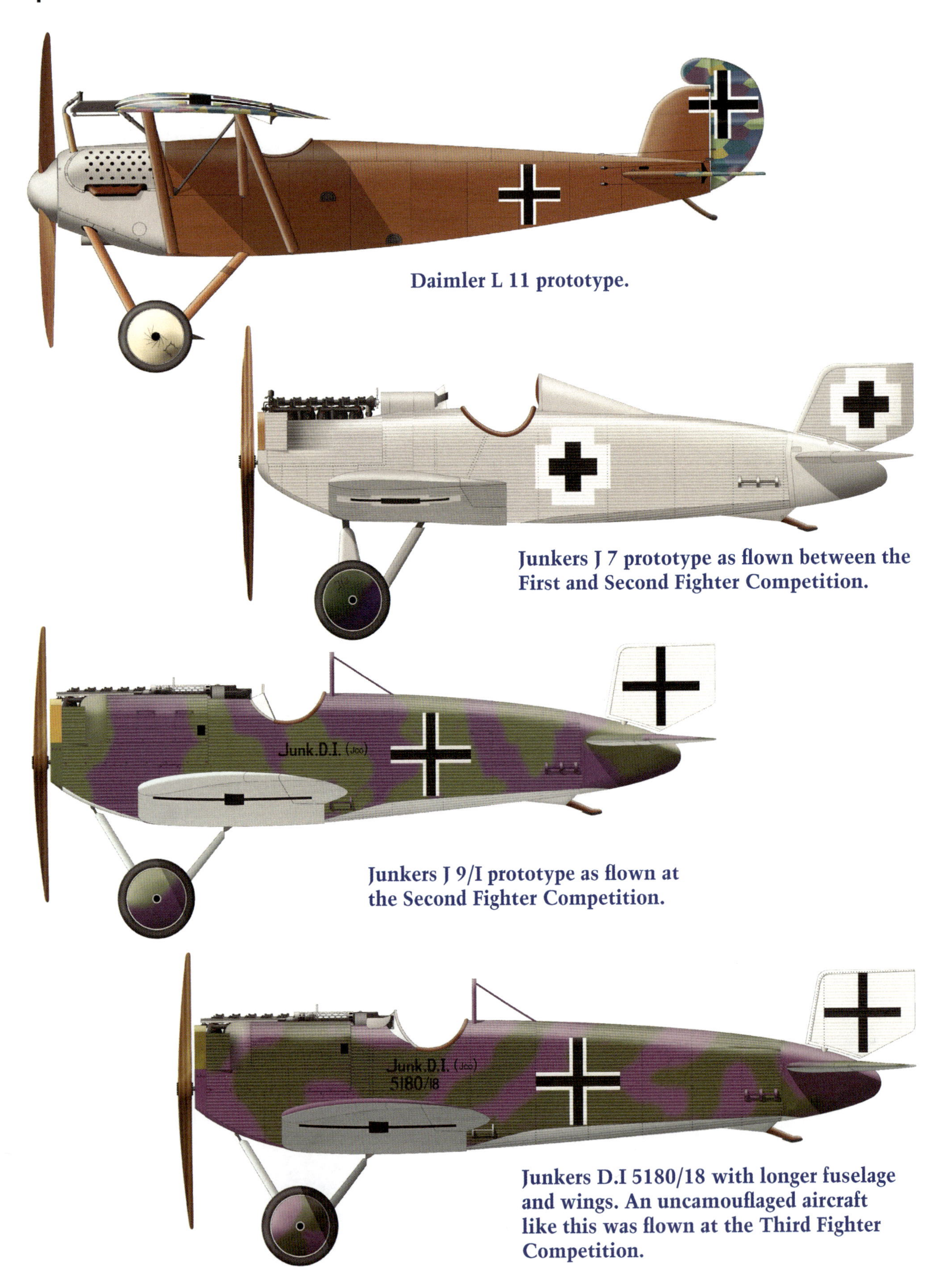

Daimler L 11 prototype.

Junkers J 7 prototype as flown between the First and Second Fighter Competition.

Junkers J 9/I prototype as flown at the Second Fighter Competition.

Junkers D.I 5180/18 with longer fuselage and wings. An uncamouflaged aircraft like this was flown at the Third Fighter Competition.

Above & Below: The largely unknown Daimler L11 fighter, which would have become the Daimler D.III had the war lasted long enough for *Idflieg* to accept the aircraft, is shown here, in initial form below and with modified ailerons above. Powered by Daimler's 185 hp Mercedes D.IIIbm geared V-8, the L11 demonstrated exceptional performance, including a maximum level speed of 240 km/h and climb to 5,000 meters in 13 minutes, excellent for any aircraft and especially impressive for one using a water-cooled engine. The Daimler L11 was faster than the Rumpler D.I and had similar climb and ceiling, and may well have filled the role the troubled Rumpler D.I was intended for had the war lasted into 1919.

In Retrospect

From summer 1917 onward was a trying time for German fighter pilots because German fighter design had stagnated and they were always outnumbered by Allied aircraft, many of which were technically superior.

The German dilemma was that, although the Albatros fighters had established superiority over Allied designs in late 1916, the Albatros D.V that arrived in June 1917 was no real improvement over its predecessors; in fact, lightening its structure to improve performance weakened it, leading to failures of the lower wing spar. The wing was strong enough statically; the failure mechanism was flutter due to the lower wing spar being too far aft in the airfoil and insufficiently stiff in torsion. Flutter was not understood at the time and various modifications reduced the problem but did not solve it.

A number of competing prototypes using the same engine and armament and similar technology to the Albatros and Roland fighters failed to reach production. The Pfalz D.III was the only such design that did reach production, but other than being structurally sounder than the Albatros due to is two-spar lower wing, unsurprisingly it offered no noticeable performance improvement.

A key problem was stagnation of German aero-engine design. Once the excellent 160 hp Mercedes D.III was established in mass production, followed by the dramatic impact of the Albatros fighters, *Idflieg* was content to mass-produce the Albatros fighters and the Mercedes engine that powered them.

Fearing disruption of engine production by introduction of a new type, *Idflieg* failed to push engine development with enough vigor, with the result that the next generation of fighter designs had to use existing engines. This was a serious mistake; the engine is the heart of an aircraft and determines its performance potential. However, instead of developing new engines, which takes extensive time and resources that were limited in wartime Germany, *Idflieg* instead focused on advanced structures and aerodynamics to achieve better performance.

One result was the end of the Triplane Craze that had afflicted *Idflieg,* and by extension the German aviation industry, for so much of 1917. Although two Fokker Triplanes flew at the First Fighter Competition as examples of fighters then in production, the DFW T 34-II triplane was the only *new* triplane fighter design ever submitted for a fighter competition. Apart from that DFW and the Naglo D.II, a retrograde quadraplane design, all other designs for the fighter competitions were either biplanes or monoplanes, with monoplanes becoming more numerous and prominent as 1918 progressed. In fact, monoplanes won the last two competitions (along with the Rumpler D.I biplane).

In contrast, the resource-rich Allies pursued engine development to increase power for better performance and relied on existing structural methods and materials to facilitate mass production.

The Allies fielded aero-engines of many configurations, not only rotary engines but water-cooled radial engines and both air and water-cooled V-8 and V-12 engines. Shockingly from the German standpoint, no German warplane flew an operational mission in WW1 powered by a German V-8 or V-12 engine. About 90% of German aero-engines produced were water-cooled, inline six-cylinder engines; the remaining 10% were rotaries.

Left: The Zeppelin D.I, with its all-metal stressed-skin construction, previewed the combat aircraft of the next great war to come, although most of those aircraft would be cantilever monoplanes. The Zeppelin D.I had the most sophisticated structure of any WW1 fighter.

Even after *Idflieg* directed the German aero-engine manufacturers to develop light-weight V-8 engines for fighters, the companies were unable to deliver production V-8 engines in quantity before the armistice. This failure was despite prolonged development and the ready availability of captured Hispano-Suiza V-8 engines for reference. Similarly, the excellent line of Rolls-Royce V-12 engines were never matched in Germany. And while the American Liberty Plane was criticized for producing an airframe from 1916 during 1918, the V-12 Liberty engine that powered it was very successful and Germany had nothing comparable to it.

Different Design Philosophies

A good example of the difference in the German and Allied design approaches is the contrast between the last fighters introduced into combat before the armistice. Germany introduced the Fokker D.VIII, a parasol monoplane that was the minimum practical fighter. It had an innovative, cantilever wooden wing and welded steel tube fuselage and tail covered by fabric. These materials were readily available even in Germany, especially the plywood for the wing. reaching combat in August 1918, it was powered by a rotary engine nearly identical in power to that used in the Fokker E.III in the summer of 1915! The D.VIII thus combined a 1915 vintage engine with 1918 structural and aerodynamic designs.

In contrast the Sopwith Snipe, a larger, more powerful derivative of the 1916 Camel, was powered by a new rotary engine. The Snipe combined a 1918 engine with 1916 structural design and aerodynamics. Snipe development was prolonged in a futile attempt to match the handling characteristics of the new Fokker D.VII. While the Snipe offered good performance due to its powerful new engine, its thin wing precluded it from matching the lift and benign stall qualities of the thick-wing Fokker. Furthermore, the additional weight and drag of the Snipe's two-bay wing limited its performance advantage over the older Sopwith Camel. Interestingly, the D.VIII and Snipe were competitive in speed, with the lighter D.VIII having superior climb despite having less than half the power (110 hp vs. 230 hp).

Left: This unarmed Fokker D.VIII, undergoing post-war performance tests at McCook field in the US, represented the minimum practical fighter aircraft for the time. Its sophisticated wing and simple steel-tube fuselage were combined with a low-power engine, exemplifying the German focus on structures and aerodynamics for performance.

Right: The Sopwith Snipe was the last new British fighter to arrive at the front. The Snipe combined a powerful, 230 hp rotary engine introduced in 1918 with structural and aerodynamic technology essentially the same as 1916 Sopwith designs, exemplifying the Allied focus on more powerful engines for improved performance.

Bibliography

Books

Franks, Norman, Bailey, Frank, and Guest, Russel, *Above the Lines*, Grub Street, 1993.

Franks, Norman, Bailey, Frank, and Duiven, Rick, *The Jasta Pilots*, Grub Street, 1996.

Franks, Norman, and VanWyngarden, Greg, *Fokker D.VII Aces of World War I, Part I*, Osprey, 2003.

Gray, Peter, and Thetford, Owen, *German Aircraft of the First World War*, second revised edition, New York: Doubleday & Company, Inc., 1970.

Green, William, and Swanborough, Gordon, *The Complete Book of Fighters*, Smithmark Publishers, 1994.

Gray, Peter, and Thetford, Owen, *German Aircraft of the First World War*, second revised edition, New York: Doubleday & Company, Inc., 1970.

Herris, Jack, *Development of German Warplanes in WWI*, Aeronaut Books, 2012.

Herris, Jack, *Germany's Triplane Craze*, Aeronaut Books, 2013.

Herris, Jack, *Pfalz Aircraft of WWI*, Aeronaut Books, 2012.

Herris, Jack, and Pearson, Bob, *Aircraft of World War I: 1914–1918*, London: Amber Books Ltd., 2010.

Toepfer, Karl E., *Empire of Ecstacy: Nudity and Movement in German Body Culture, 1910–1935*, University of California Press, 1997.

Articles

Grosz, Peter M., "Kondor's Tiny Fighters" *Windsock* Vol.13, No.2, March/April 1997, p.11–20.

Grosz, Peter M., "Kondor's D.VI Fighter" *Windsock* Vol.13, No.5, September/October 1997, p.24–27.

Grosz, Peter M., "Frontbestand" *WW1 Aero* No.107, Dec. 1985, p.60–66.

Grosz, Peter M., "Frontbestand" *WW1 Aero* No.108, Feb. 1986, p.66–69.

Grosz, Peter M., "Daimler Aircraft 1914–1919" *Over the Front* No.21-3, Fall 2006, p.250–280.

Reddehase, Erwin, "German and Austrian Aircraft Engines of the First World War", *Cross & Cockade*, Vol.5 No.4, Winter 1964, p.321–326.

Above & Left: The all-metal Zeppelin D.I fighter was powered by either an over-compressed Mercedes or 185 hp BMW.IIIa engine. The D.I above was tested in the U.S. postwar. The fuel tank was suspended below the fuselage and was planned to be jettisonable in case of fire.

Daimler L11

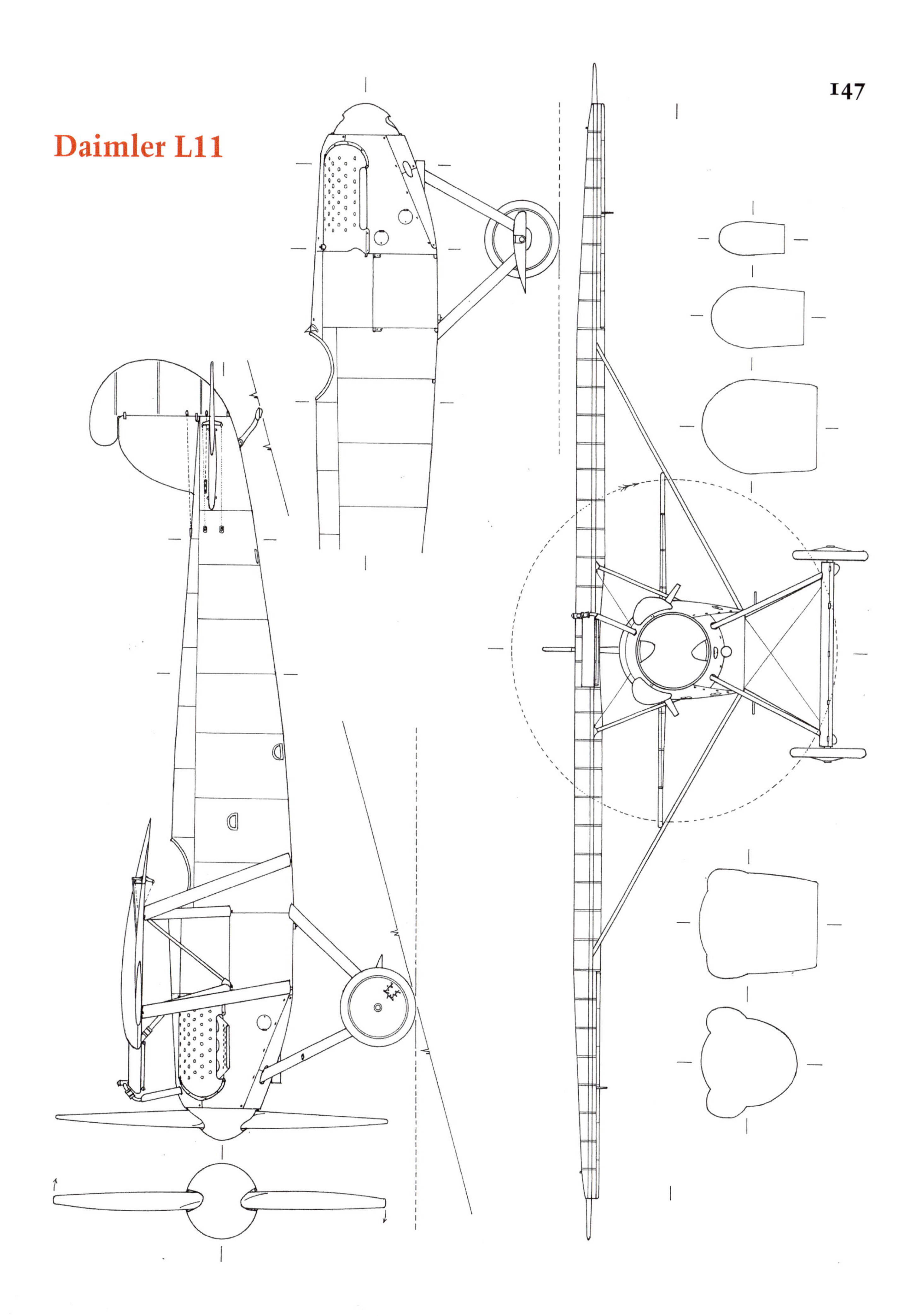

Daimler L11

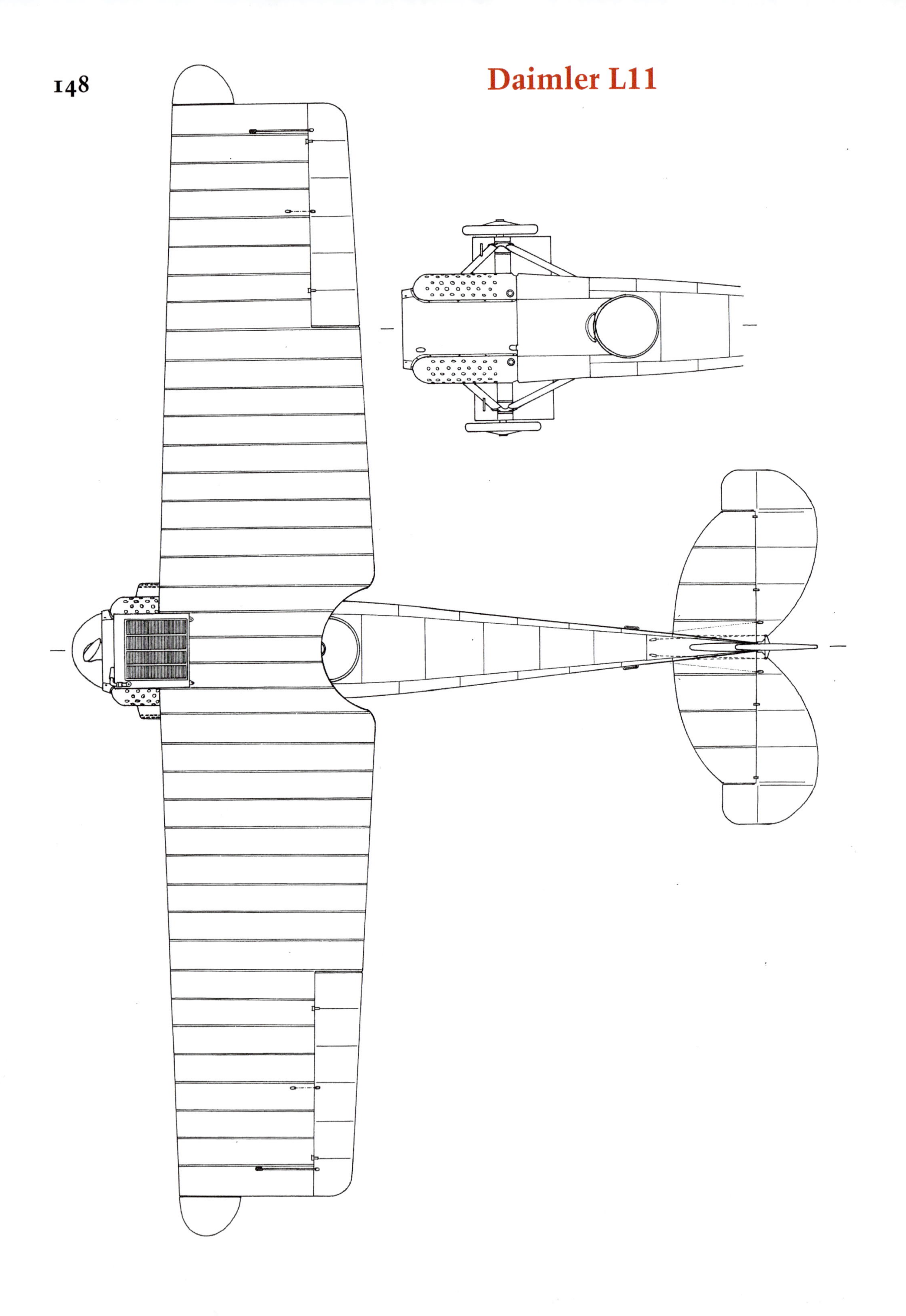

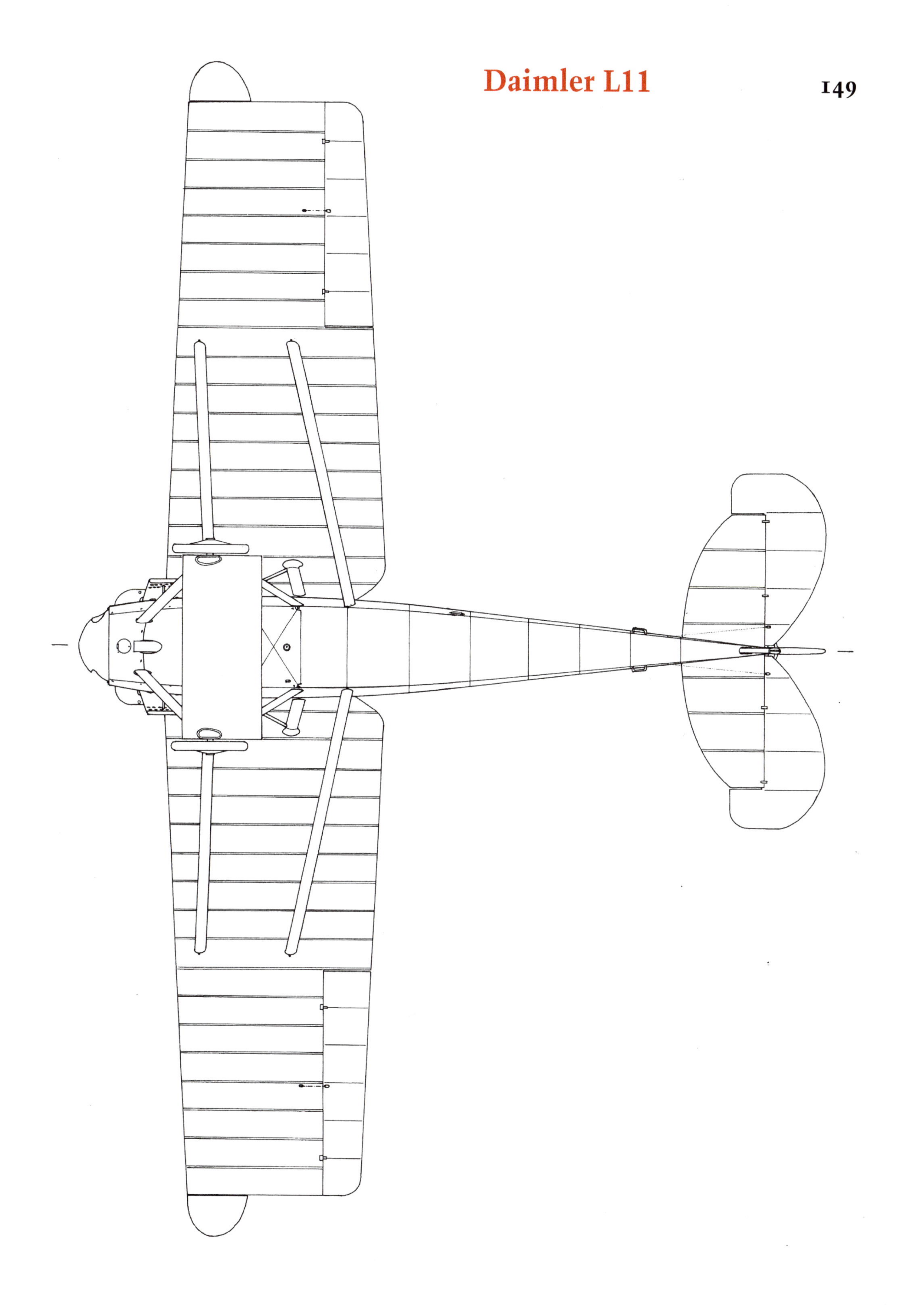

Kondor D.II

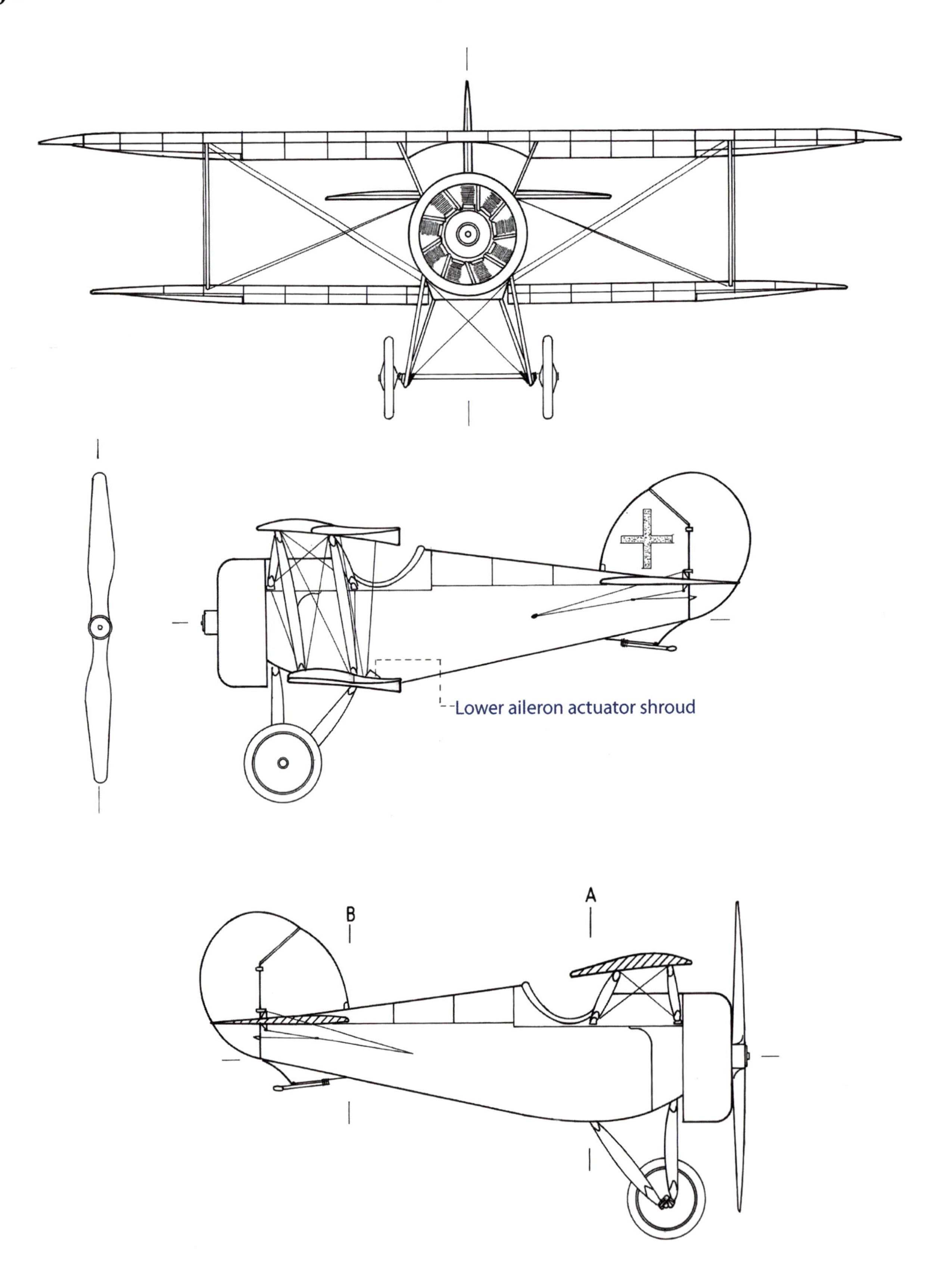

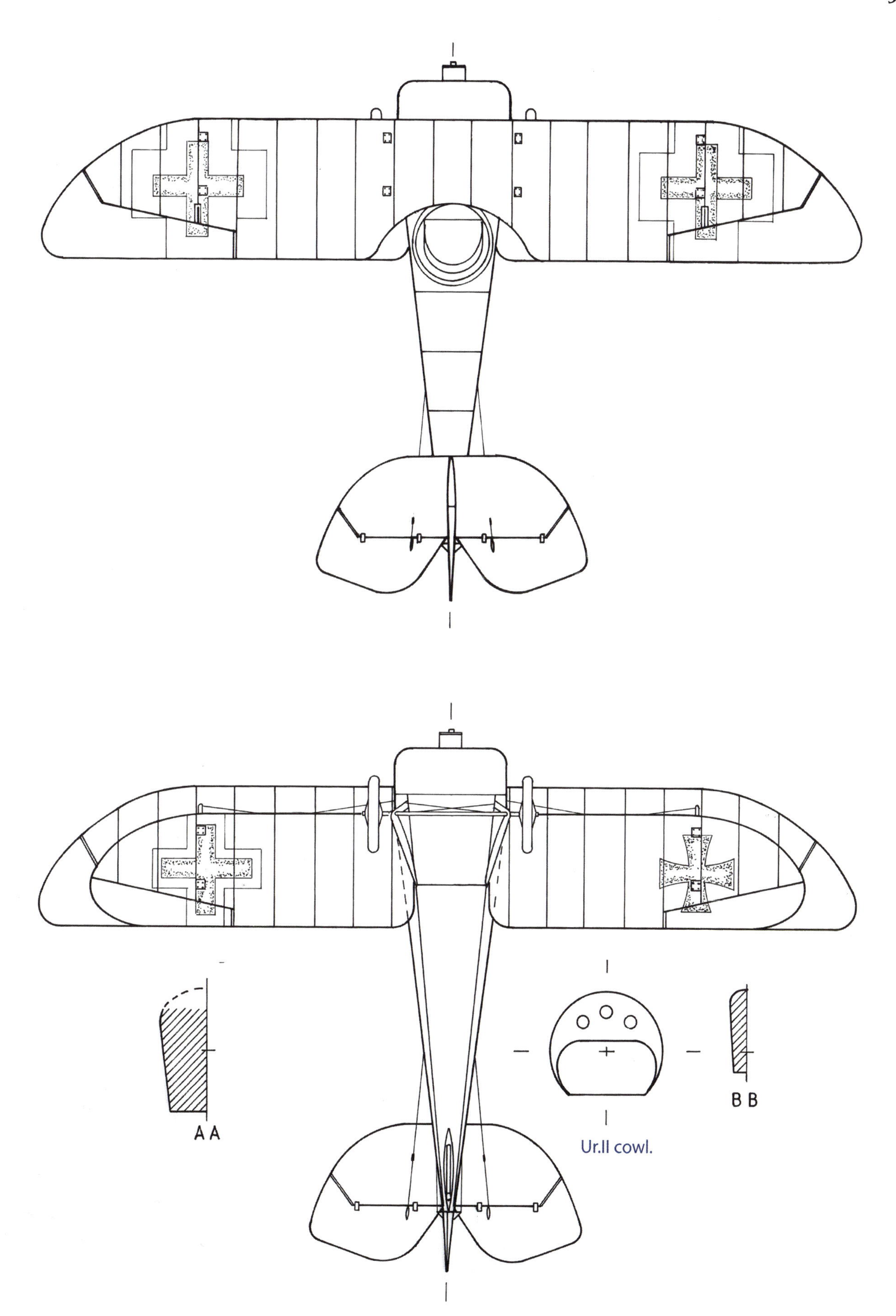
AA
BB
Ur.II cowl.

Kondor D.VI

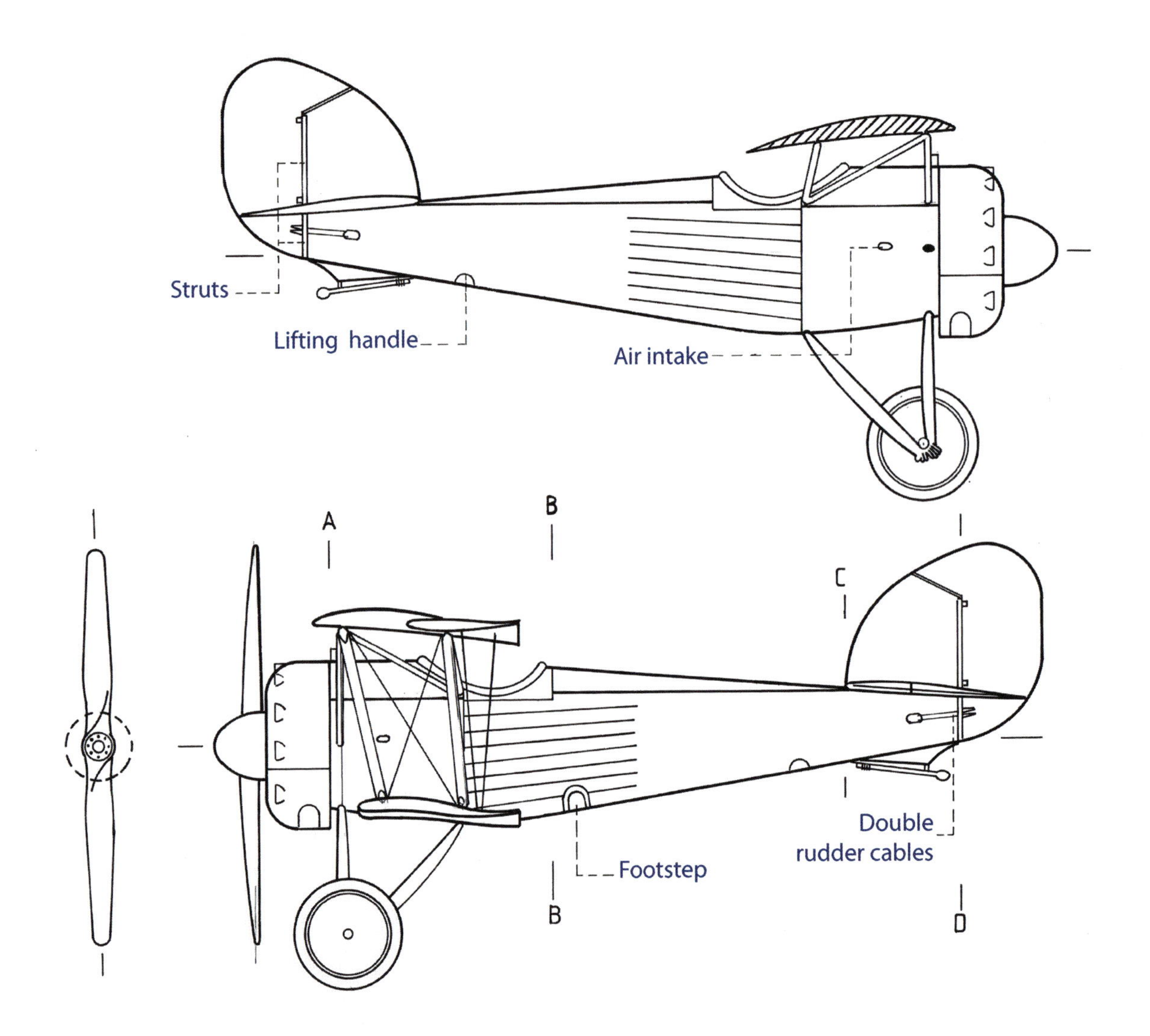
Struts
Lifting handle
Air intake
A
B
C
Double
rudder cables
Footstep
B
D

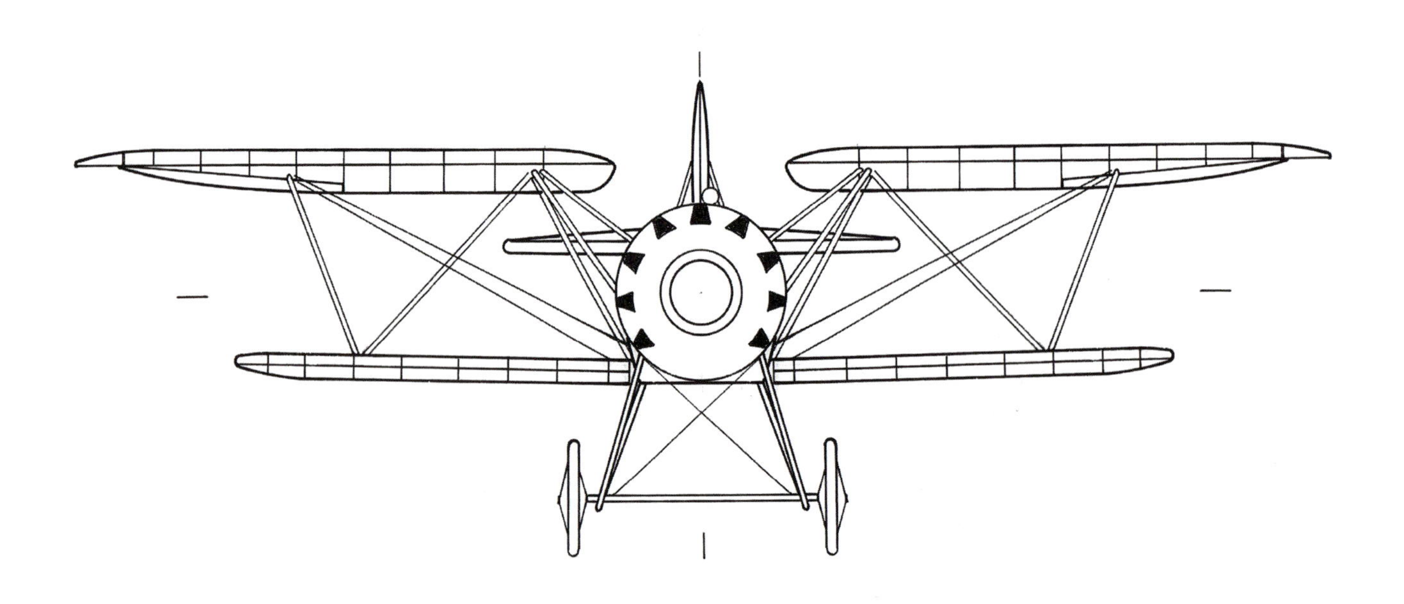

BB

CC

Kondor D.VI

AA

D D

Kondor E.III & E.IIIa

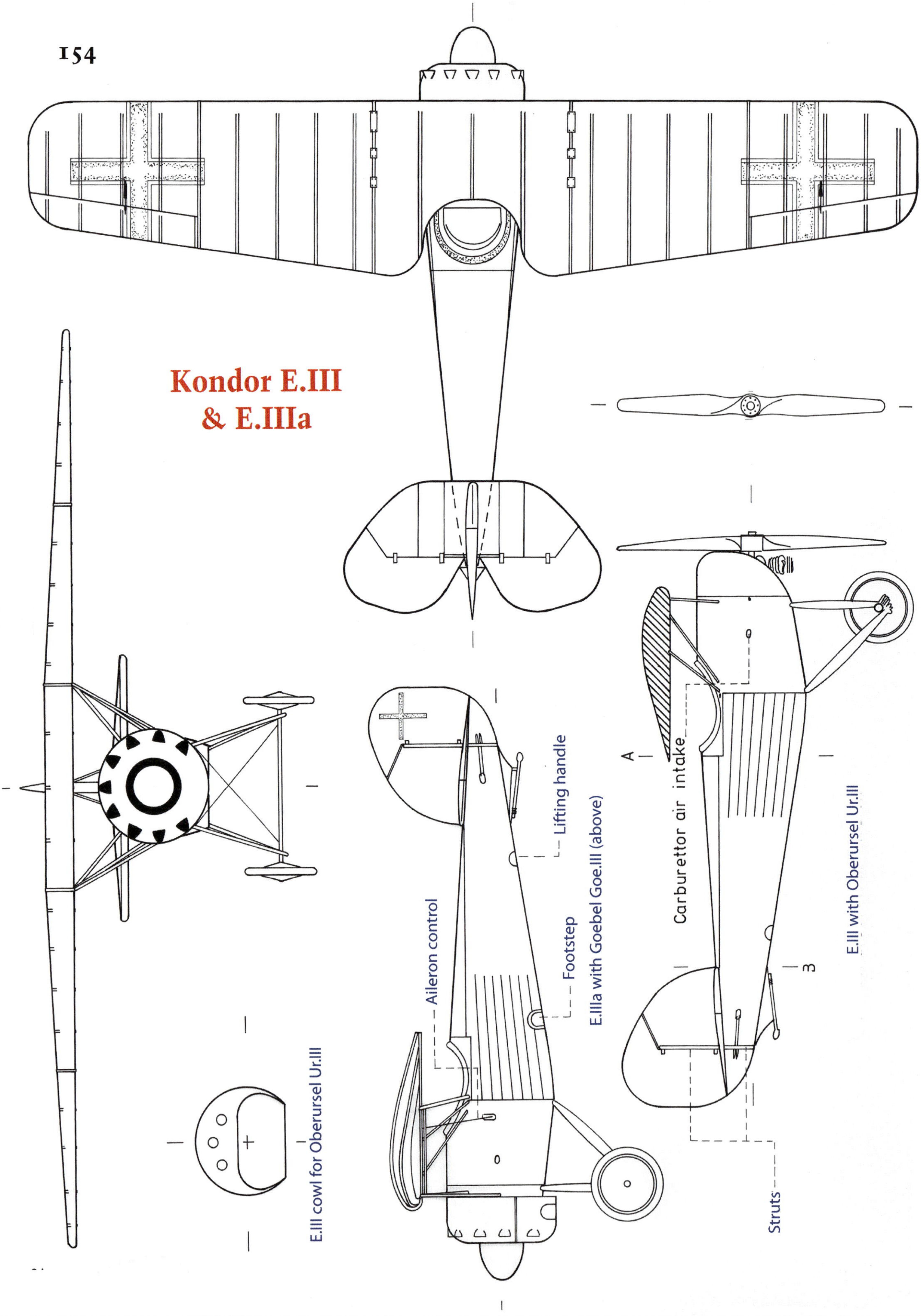

E.III cowl for Oberursel Ur.III

Aileron control

Footstep

E.IIIa with Goebel Goe.III (above)

Lifting handle

Carburettor air intake

A

3

Struts

E.III with Oberursel Ur.III

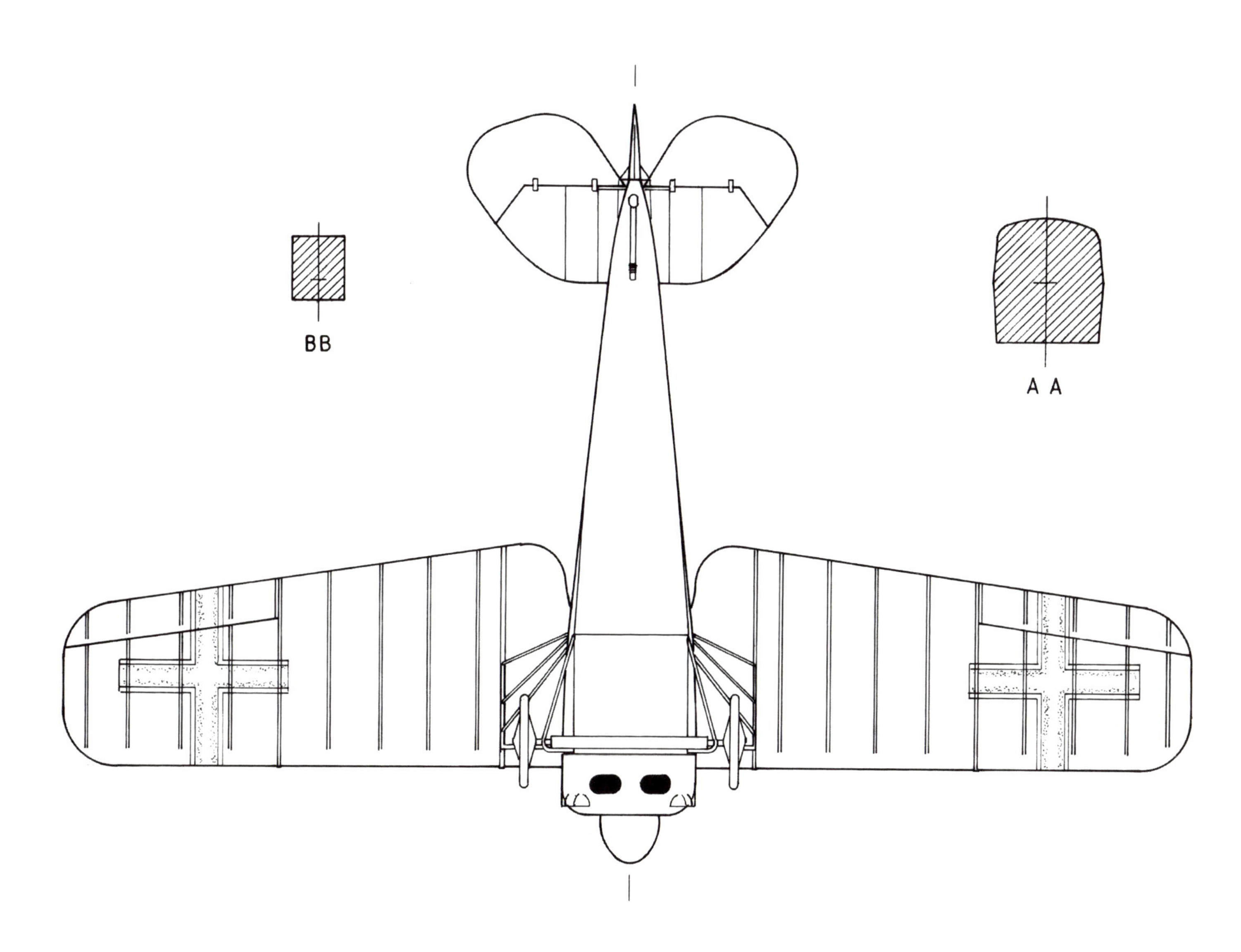
BB
A A

SSW D.VI

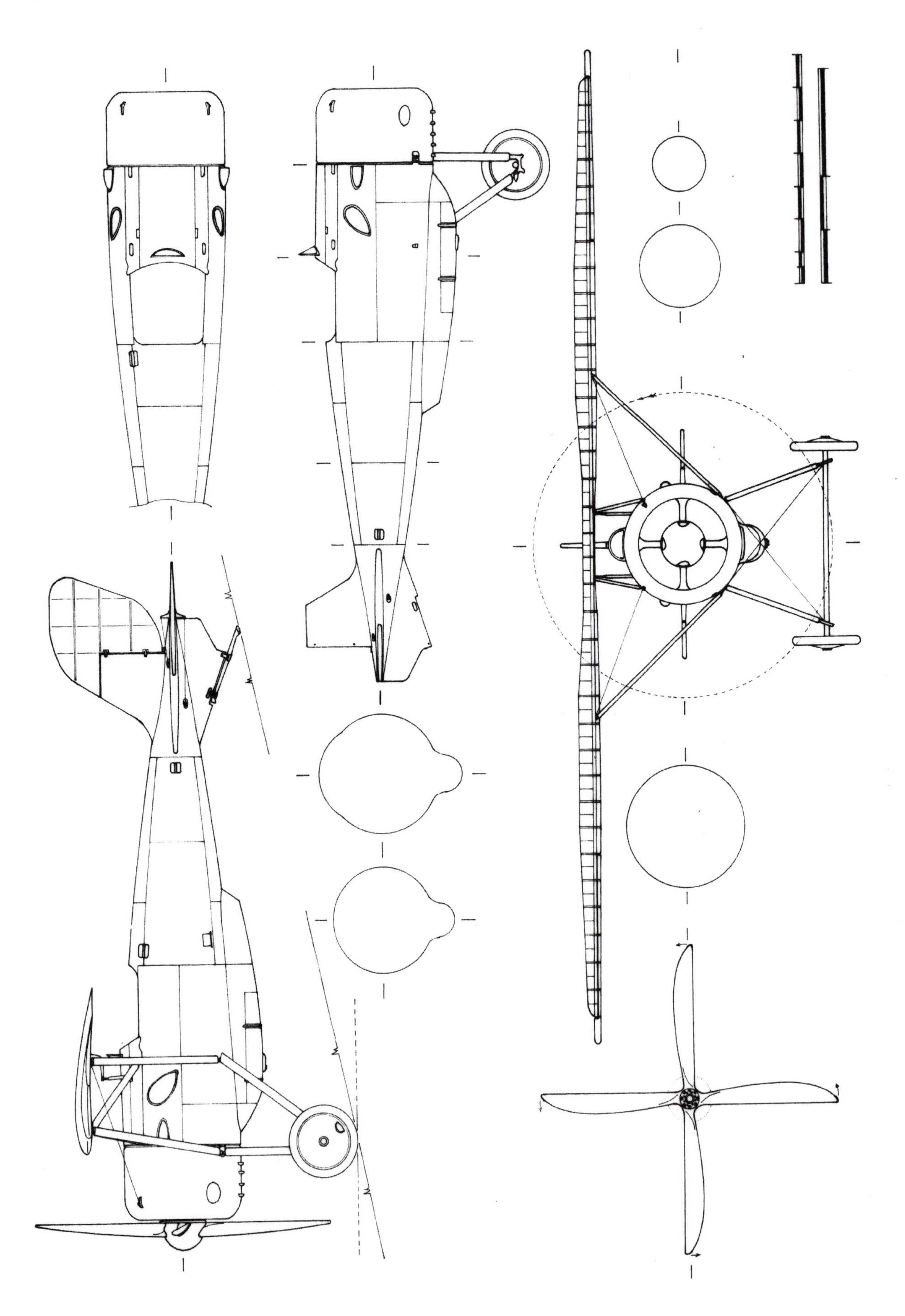

SSW D.VI

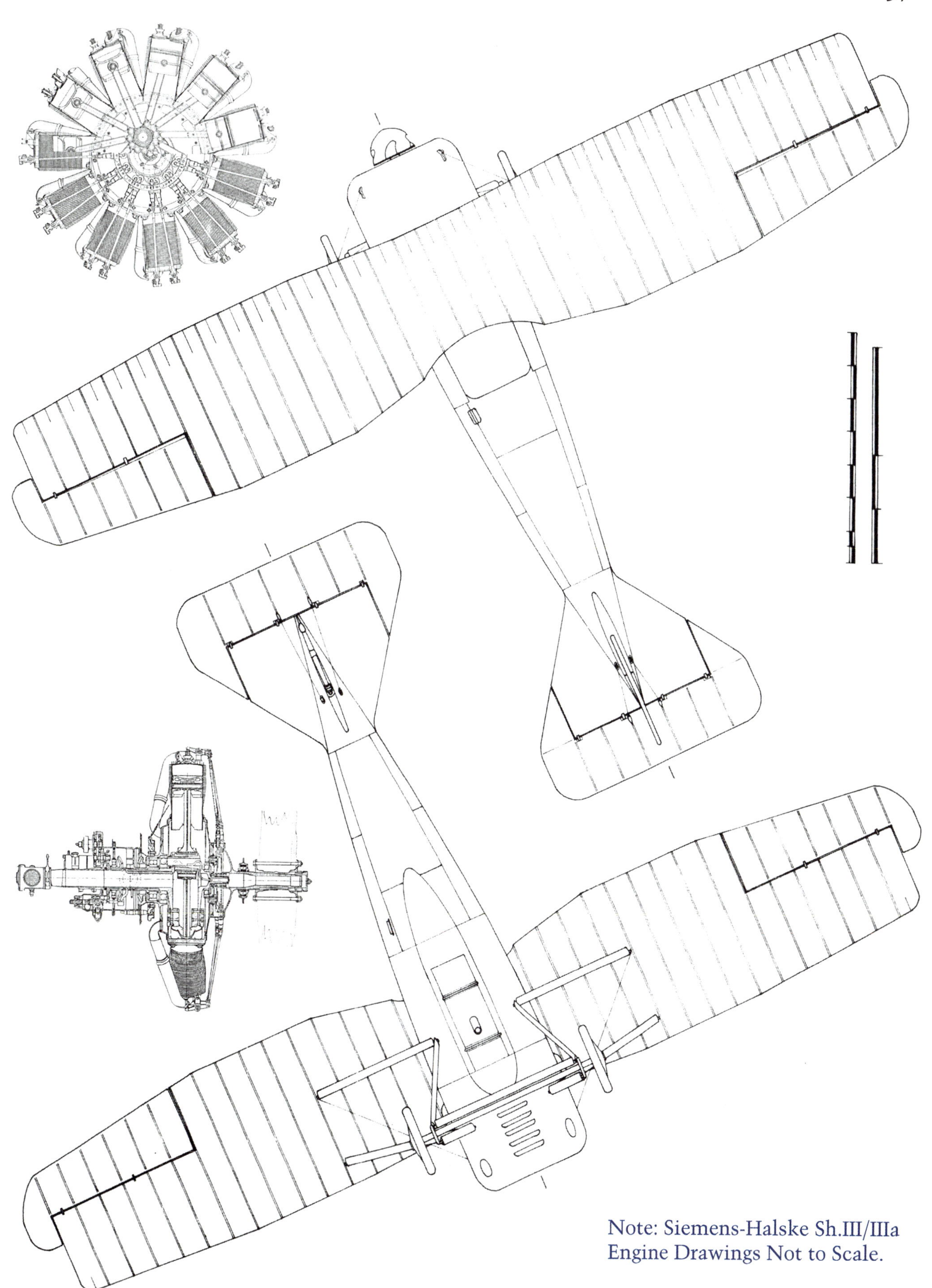

Note: Siemens-Halske Sh.III/IIIa Engine Drawings Not to Scale.

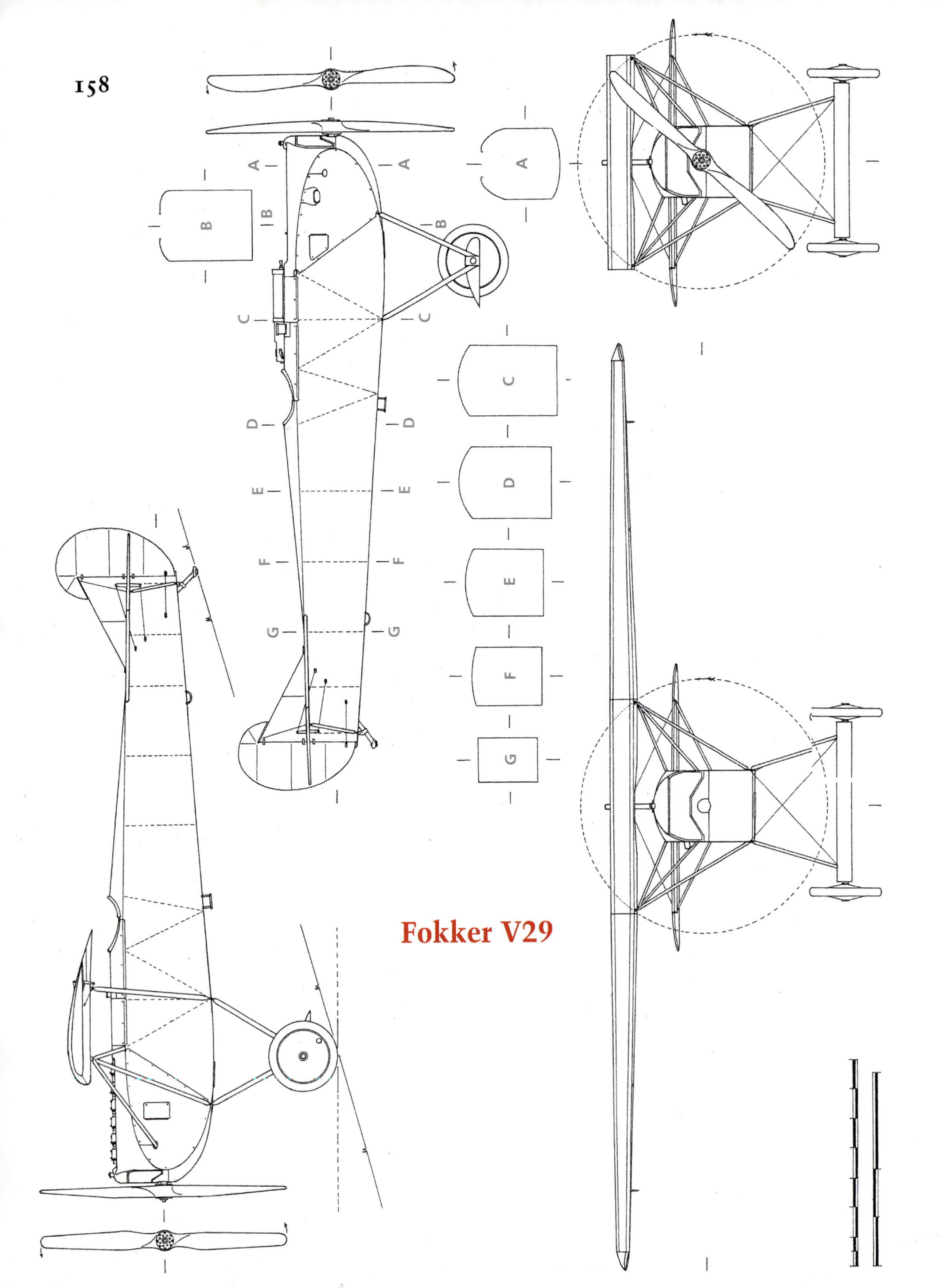
A
B
C
D
E
F
G
Fokker V29

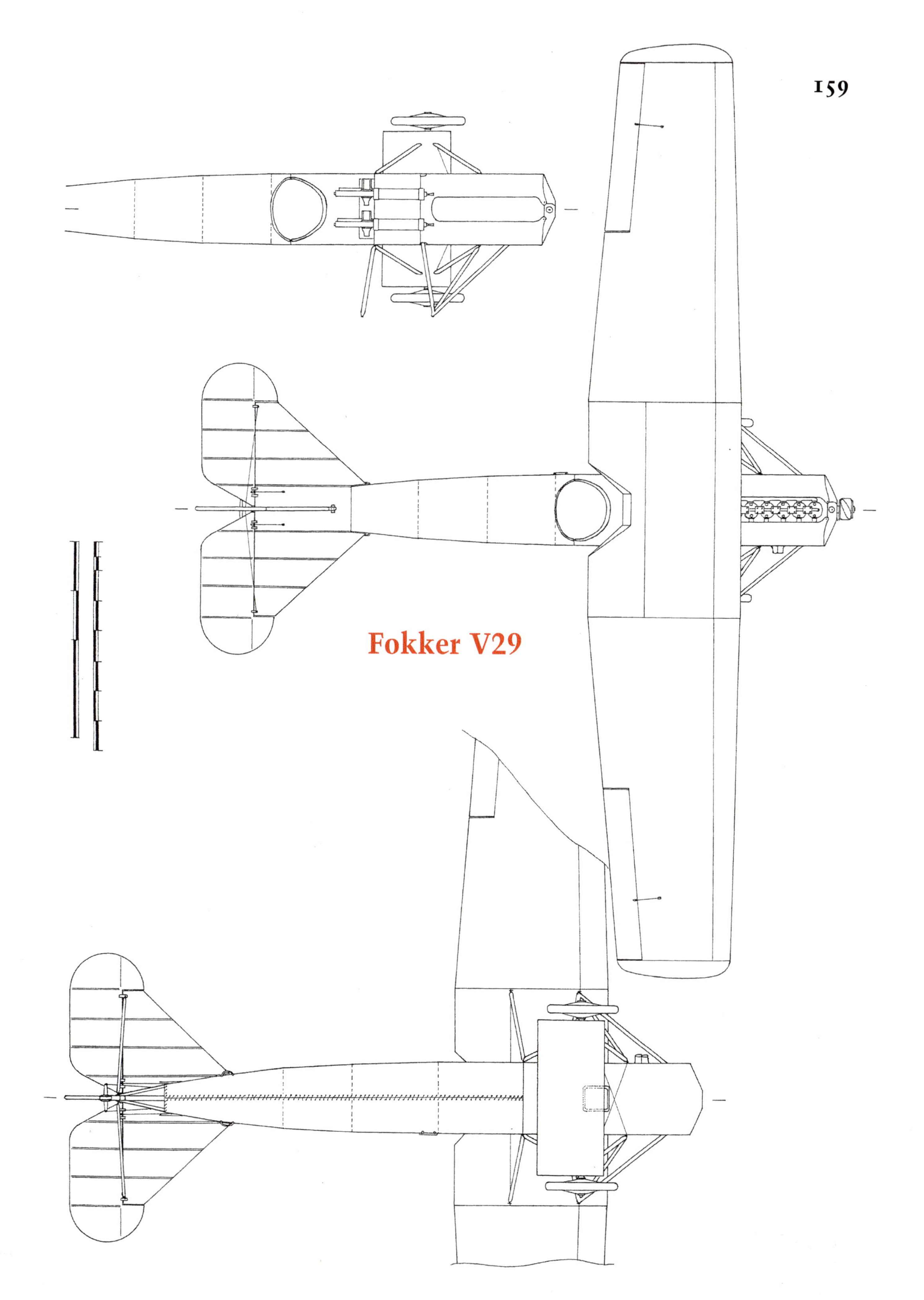

Fokker V29

Zeppelin D.I

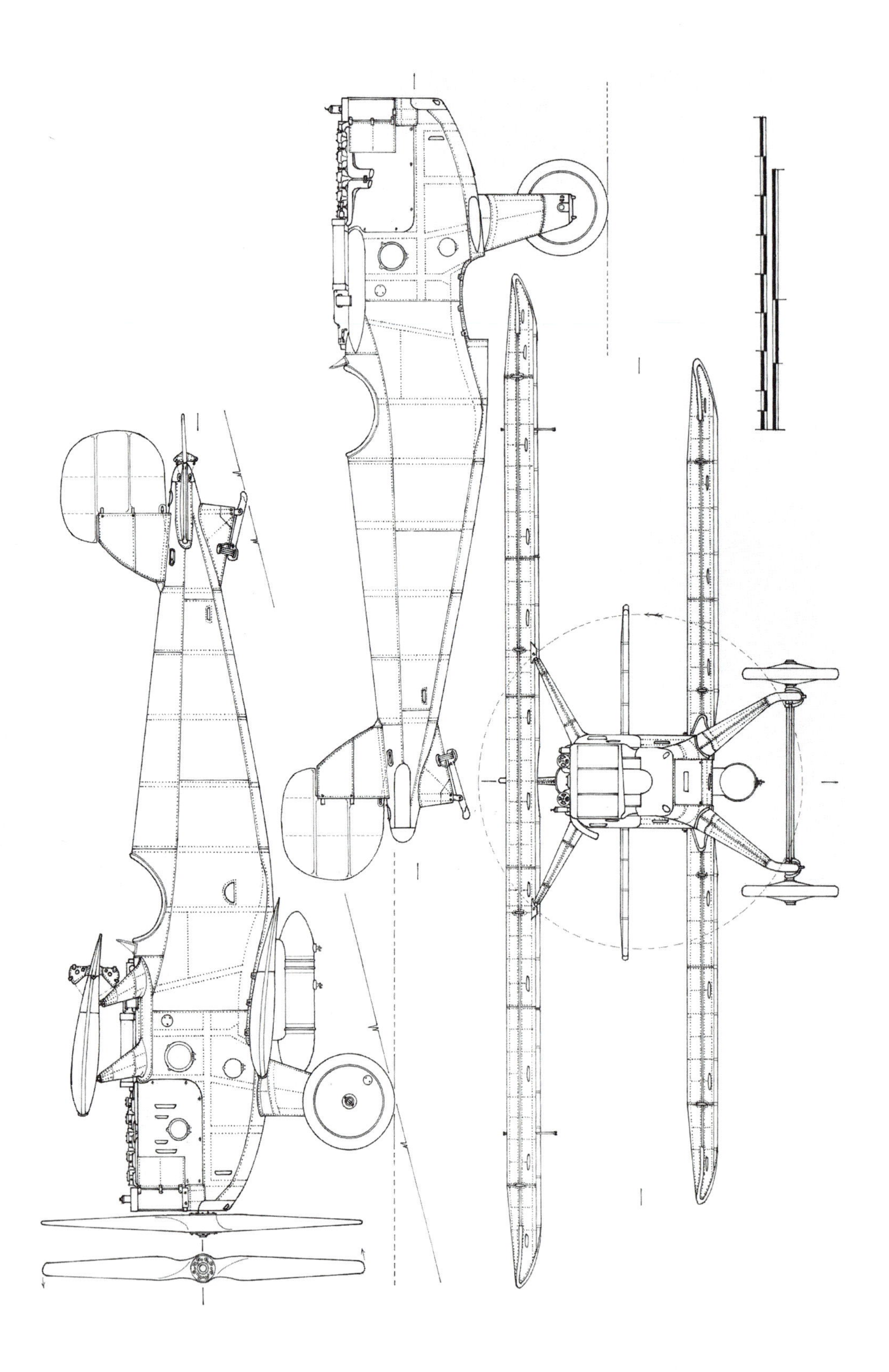

Zeppelin D.I

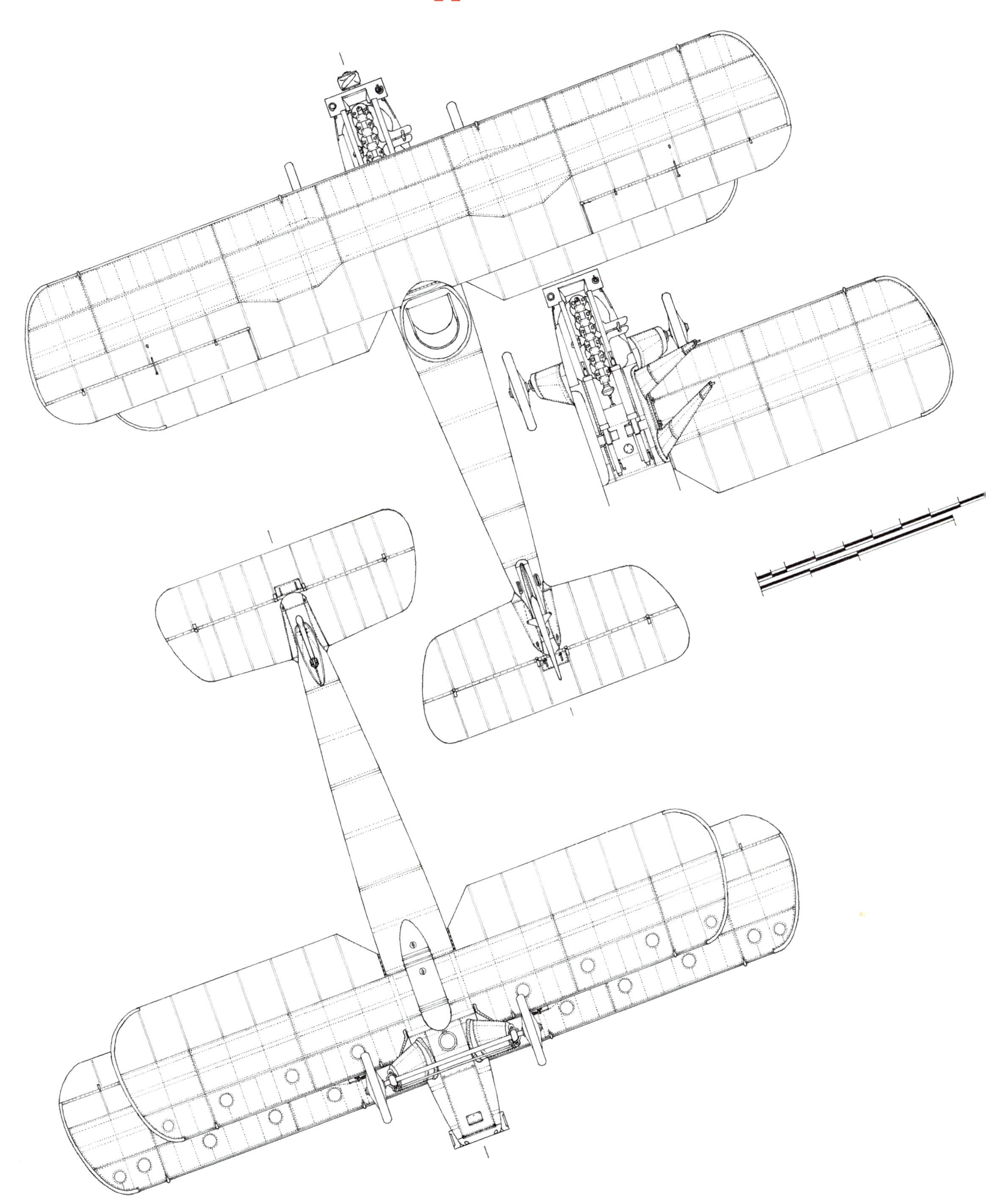

Aviatik D.VI

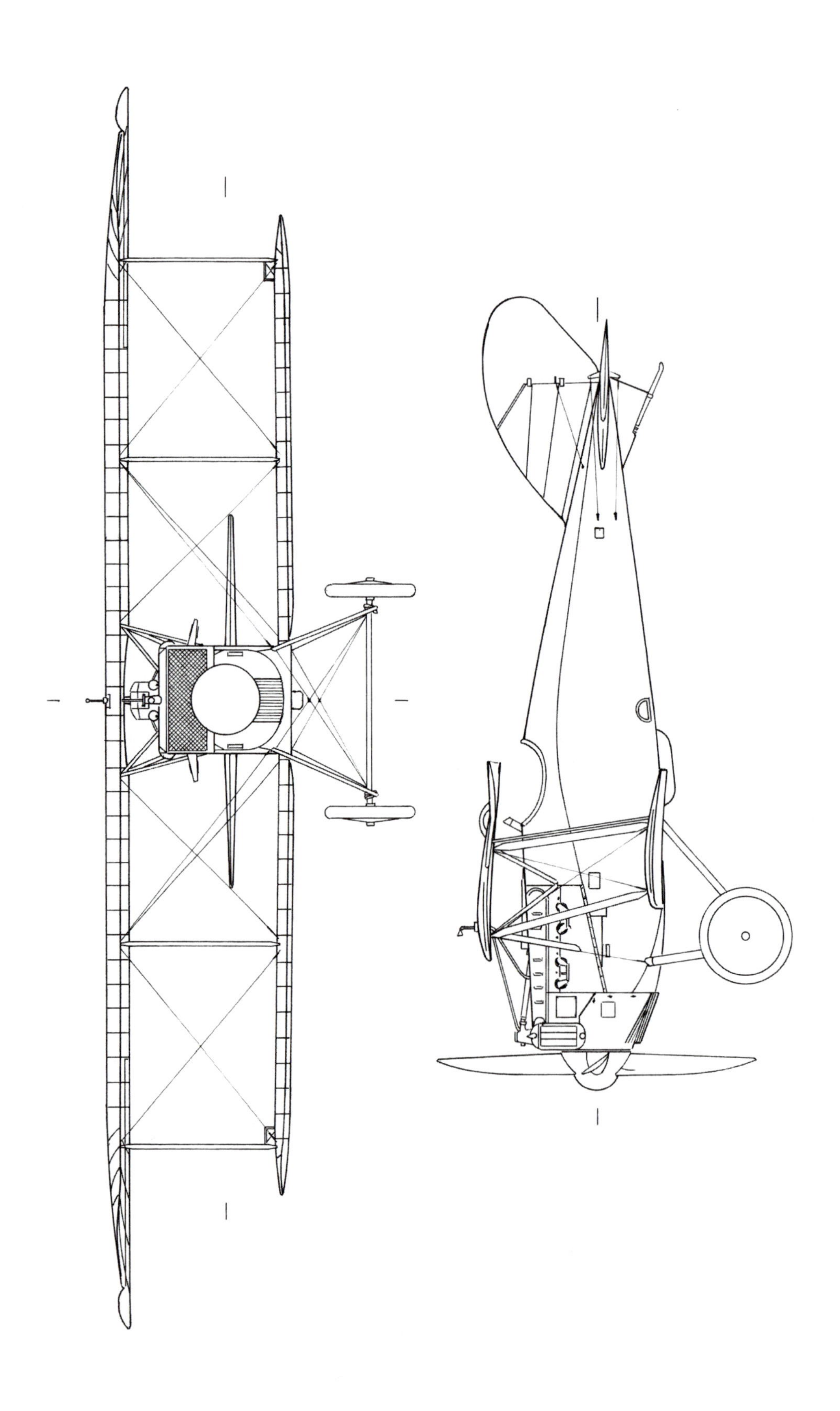

Aviatik D.VI

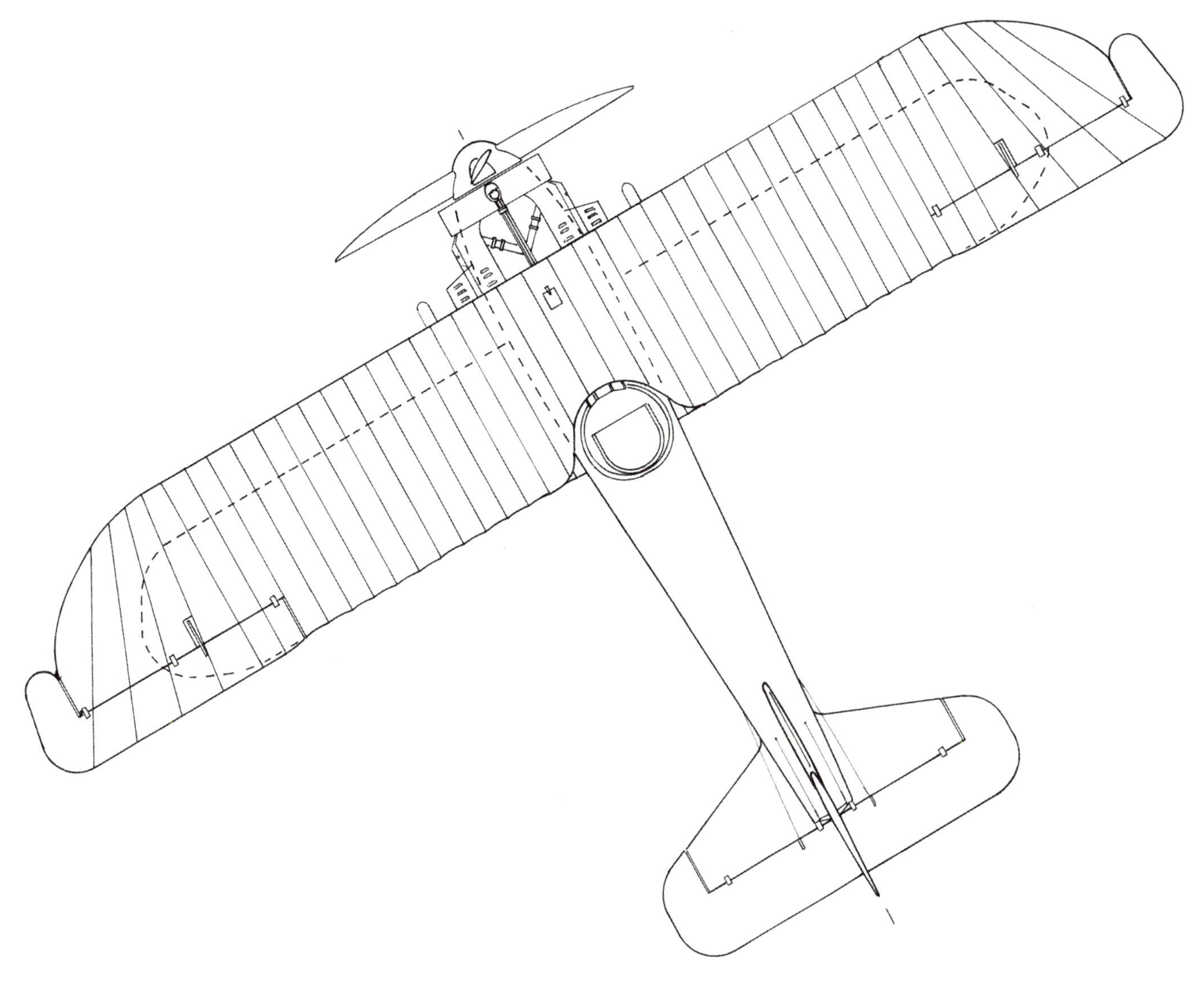

Index

Aviatik D.VII

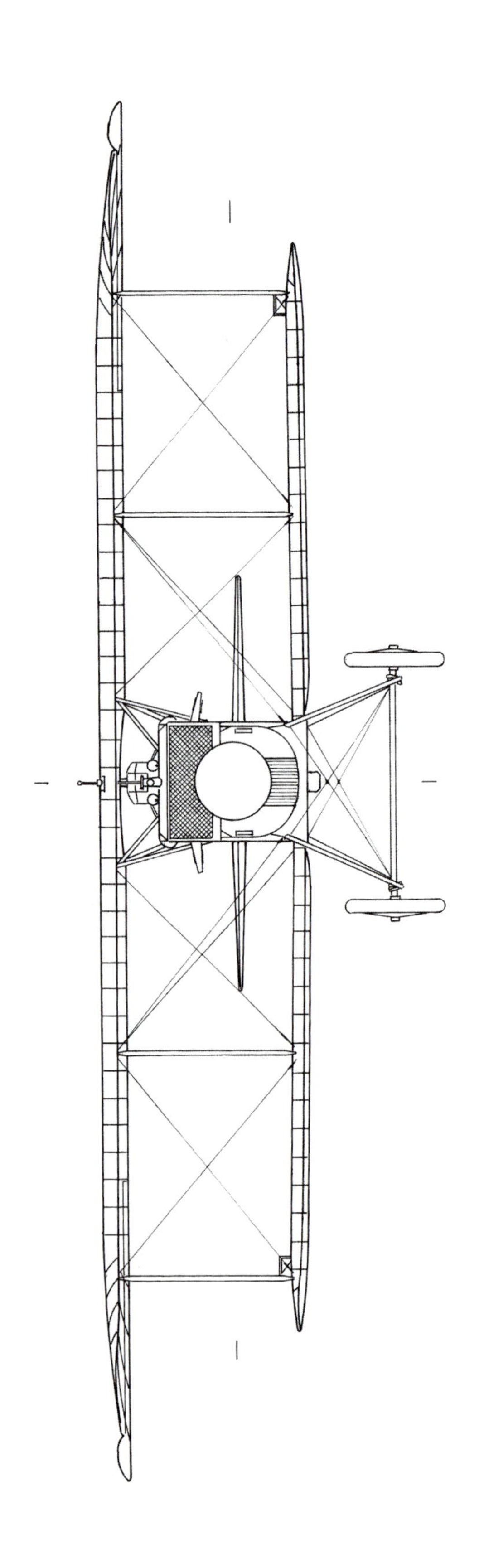

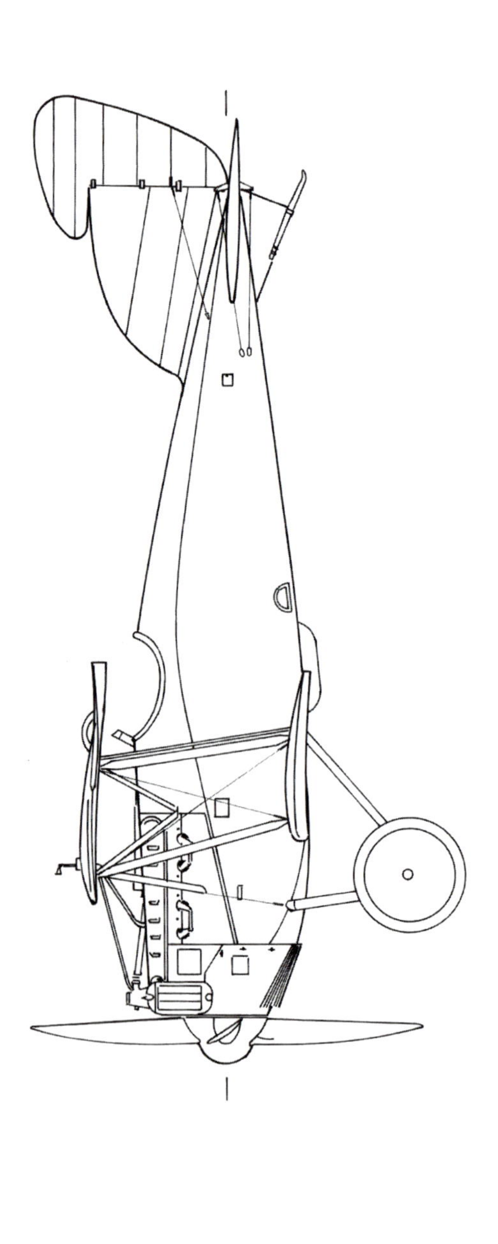

Aviatik D.VII

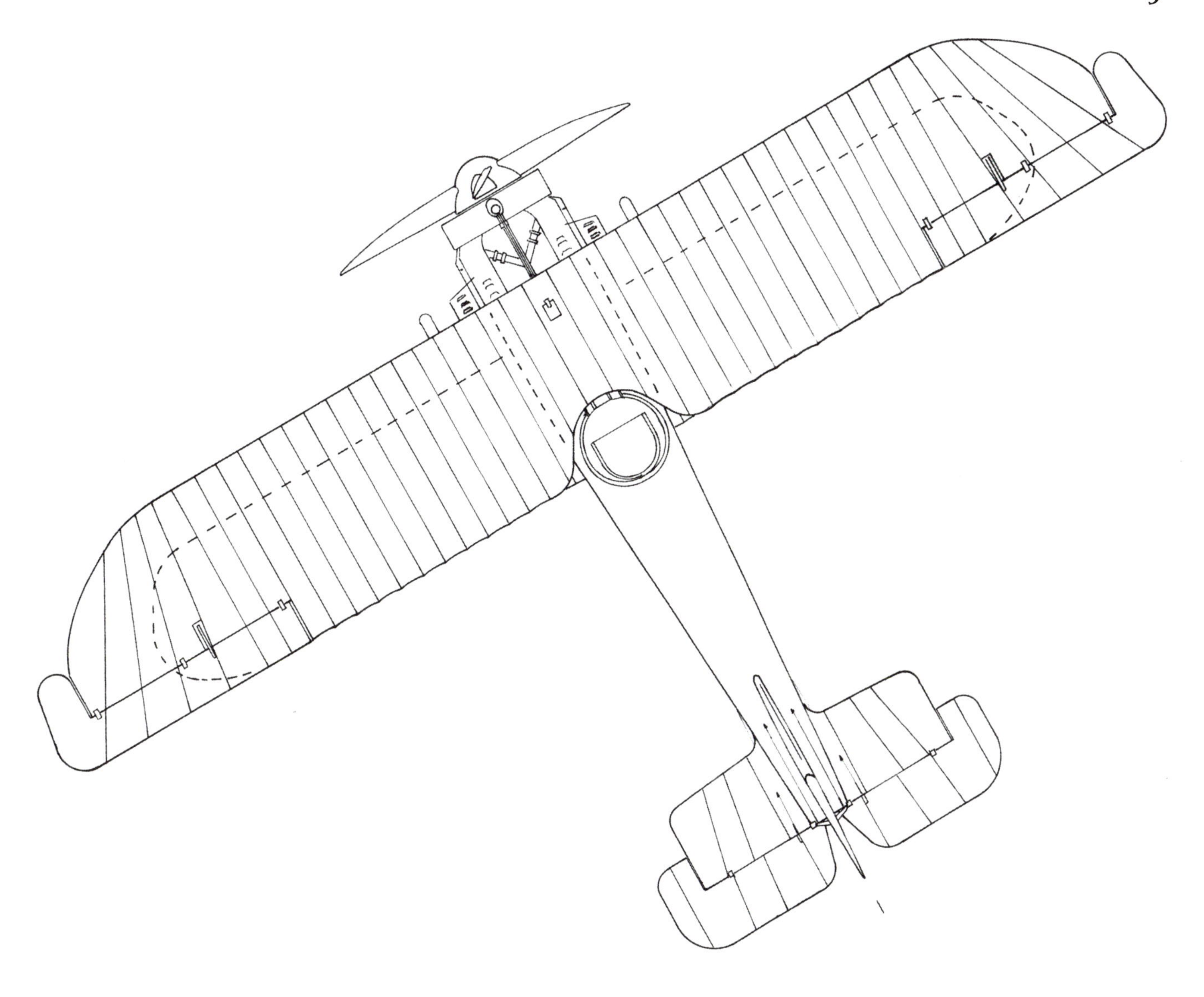

Index (Continued)

Afterword

Above: This engaging portrait shows the leaders of the Flying Circus; from left, *Lt.* Curt Würsthoff (*Jasta* 4), *Oblt.* Wilhelm Reinhard (*Jasta* 6), *Rittmeister* Manfred von Richthofen (*JGI*), *Lt.* Erich Löwenhardt (*Jasta* 10), and *Lt.* Lothar von Richthofen (*Jasta* 11). The three middle *Pour le Mérite* awards were drawn in; Reinhard was nominated for the award but was killed before he received it; Löwenhardt did not receive his until after this photo was taken; and the *Rittmeister* was not wearing his.

The story of the German fighter competitions of 1918 is a key part of the story of 1918 German fighter development, and we have enjoyed telling part of the broader story as well as providing details of the various competitions and the competing aircraft. We trust you have enjoyed the story as well.

And if the stars of the story are the airplanes, the men are essential. The reason the airplanes were built was to provide the most effective 'tools of the trade' for gentlemen like those illustrated above. Not only did these men use the airplanes, during the fighter competitions they helped develop the airplanes and decide which ones were going to be produced for them and for their comrades.

So the story of the fighter competitions is a human story as well as an airplane story. And a single book can tell only a very small part of that greater human story. We invite you to explore 'the rest of the story' in the many other books published by Aeronaut Books, such as our "*Aviation Art of...*" and *Blue Max Airmen* series among others.

25279319R00092

Made in the USA
Lexington, KY
18 August 2013